Building the Wireless Age - A Unique Marconi Exhibition

At The Original Marconi Wireless and Telegraphy Works in Hall St Chelmsford, the World's first wireless factory which Marconi established in 1898.

Marconi Science WorX: Chelmsford Civic Society in collaboration with BBC Essex

Open 11th March to 29th May 2016

Every Sat. & Sun.   11.00am - 3.00pm   Free entry

We are pleased to announce the following talks:

• **Friday 18th March** at 19.00 for 19.30 start. Hall St and Marconi: Building the Wireless Age **Speaker Tim Wander**, curator, Marconi historian and author.

• **Thursday 14th April** at 19.00 for 19.30 start. The Role of the Wireless in the Titanic Tragedy. **Speaker Tim Maltin**, Titanic historian, author and broadcaster, this is on the anniversary of the sinking of the Titanic

• **Friday 22nd April** at 19.00 for 19.30 start. The Battle of Jutland. **Speaker Dr Elizabeth Bruton**, University of Oxford

To book tickets, £5 each, please visit www.chelmsfordcivicsociety.eventbrite.com

If you would like any more information or you would like to volunteer for this project please

visit www.facebook.com/marconiscienceworx or email info@chelmsfordcivicsociety.co.uk.

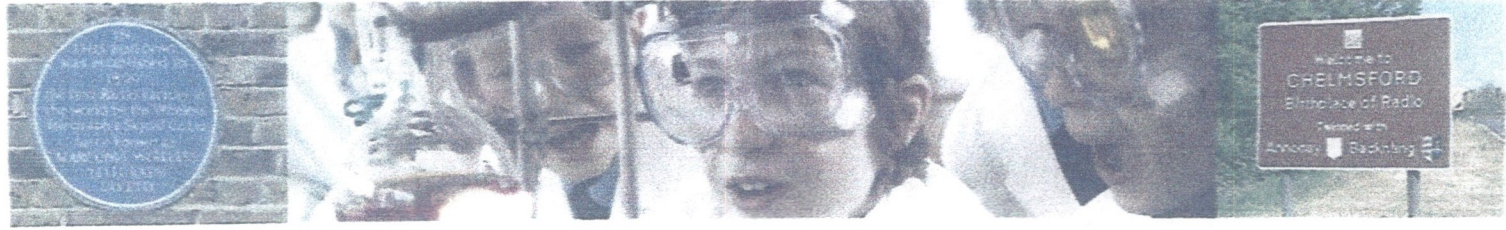

Registered charity

Marconi's Hall Street Works
1898-1912
The World's First Wireless Factory

Tim Wander

First Published by New Generation Publishing in 2016

First Edition

Originally produced as a limited edition of just 60 numbered and signed copies as part of the 'Marconi: Building the Wireless Age Exhibition' which celebrated the history and legacy of Marconi in Chelmsford. The exhibition ran from March to May 2016 in the newly refurbished Hall Street building prior to its final conversion into apartments and commercial space. The exhibition was organised and presented by the Chelmsford Civic Society (Reg. Charity # 271779) affiliated with the Marconi Heritage Group and Marconi Science Worx in collaboration with BBC Essex at the request of the building's developer, MAC Design and Build Ltd.

ISBN 978-1-78507-667-1

Parts of this text have been extracted from :

'2MT Writtle - The Birth of British Broadcasting'
by Tim Wander, Authors Online. ISBN 978-0-7552-0607- 0 © T.R. Wander 2010

'Marconi's New Street Works 1912-2012. Birthplace of the Wireless Age'
by Tim Wander, Authors Online. ISBN 978-0-7552-0693-3 © T.R. Wander 2013

'Guglielmo Marconi - Building the Wireless Age'
by Tim Wander, New Generation Publishing. ISBN 978-1-7850-7481-3 © T.R. Wander 2015

The author can always be contacted via: **www.marconibooks.co.uk**

New Generation Publishing
www.newgeneration-publishing.com

Dedicated to all the wireless men
           who served at sea,
                      on the land,
                           and in the air.

and for all the engineers and staff of the Hall Street works.

and for Stan Wood - this is the book he always wanted to write.

**Marconi's Wireless Telegraph Company Ltd
Hall Street works**

By the same author

2MT Writtle - The Birth of British Broadcasting (1988)

Marconi on the Isle of Wight (2000)

2MT Writtle - The Birth of British Broadcasting (2010)

MARCONI'S NEW STREET WORKS
1912 – 2012
Birthplace of the Wireless Age (2011)

Marconi on the Isle of Wight (2013)

The Marconi Company and Writtle (2013)

Northwood House. A Guidebook, History and Tour (2015)

In Preparation

Culver Cliff at War

Oyster, Oveners and Racing Gaffes
The Paskins Boys on the Isle of Wight. 1814-1914

A Chelmsford Industrial Trail
2nd Edition

For more information on these books and new projects please see
marconibooks.co.uk

MARCONI'S HALL STREET WORKS

1898 – 1912

The World's First Wireless Factory

Tim Wander

Authors Note

The traditional histories of the Marconi Company all record that the Marconi Hall Street wireless works opened in December 1898, and that Marconi visited Chelmsford and Hall Street at the end of 1898. Recent research from Chris Neale shows that Marconi's actually took up the lease at the end of January 1899. Most sources also indicate that the site, despite a 20 year lease, was effectively abandoned, and all the machinery, stock and personnel had been transferred across to the New Street Works by June 1912.

By November 1912 the Company had been approached by Messrs F. Taylor and Co (auctioneers and estate agents) who offered to lease the works for £200 per year or buy them for £3,700, possibly on behalf of a client. However it is clear that the Hall Street wireless station, across the road from the main works with its huge 200 ft aerial mast(s), continued to operate well into the First World War, playing a vital role in the birth of wireless intelligence and signals interception. A final date for the close down of this station is not known. It is not recorded after 1919 and the best estimate is that it had been closed by the end of 1916.

Marconi's Hall Street Works -

where Marconi first started to build the Wireless Age

Between 1896 and 1898...

The young Guglielmo Marconi assembled a group of experimental techniques and laboratory curiosities and had started to develop a practical 'wire-less' communication system.

His early work was to lay the foundations for a new industrial revolution and would define a new way of doing business. He was the first of a new breed of entrepreneurs, a driven and focused pioneer who would risk everything to achieve a single, incredible, at times seemingly impossible goal.

As his business grew, the challenges he encountered demanded the development of a new era of manufacturing and a factory to meet the demands of a new industry.

At the dawn of the age of mass production techniques and manufacture he started the world's first wireless factory.

At Hall Street Marconi started to build the new age of wireless and his work became the foundation for the new world of radio, broadcasting, electronics, radar, television, computers, satellite communication and cellular mobile phones.

Guglielmo Marconi, c. 1905

6

'*My* name is Guglielmo Marconi
and I have just invented the wireless.'

Toynbee Town Hall, Saturday evening, 12th December 1896,
Announced at the end of William Preece's lecture, 'Telegraphy without Wires'

'*The* Marconi invention is the only (electric) telegraph by means of which a moving object can be kept in communication with any other moving object, or a fixed station, and therefore any one can see the great use of the invention, not only to the Royal Naval authorities, but also to the mercantile marine.

A ship fitted with Mr Marconi's apparatus can not only keep in telegraphic communication with the shore up to any reasonable distance - it has been thoroughly tested up to twenty-five miles off the shore - but ships can also, if properly equipped, be warned of approaching danger or their proximity to dangerous coasts which are fitted with the wireless apparatus.'

Henry Jameson Davis,
Chairman of the Wireless Telegraph Company, 7th October 1898

'*Marconi's* work is no small achievement. His apparatus is ridiculously simple and not costly. With the exception of the Flagstaff and 150 feet of vertical wire at each end, he can place on a small kitchen table the appliances, costing not more than £100 in all, for communicating across 30 or even 100 miles of land or water.

Prof J.A. Fleming,
'On Wireless Telegraphy', reported in the *Irish Times*, 4th April 1899

'*This* is going to be a new works and it will also be an experimental wireless station.

We have been buying all the instruments we require from different firms,
but in the future we will make them here ourselves.'

We are only yet in the experimental stage.

You must come when we have the big pole erected and the instruments at work.
We shall show you some things that will astonish you.

George Kemp, Marconi's principal assistant and lifelong friend.
Speaking to the *Essex County Standard* newspaper, 29th April 1899

S.O.S - Many Hearts Bless You Today Sir, The Worlds Debt To You Grows Fast

Tribute to Marconi, printed in Punch Magazine in 1913, after the *Titanic* disaster.

CONTENTS

ACKNOWLEDGEMENTS

As usual thanks to my long suffering wife Judith for, as always, typing above and beyond the call of duty and endless cups of tea. Her contribution is incalculable as my handwriting is truly dreadful and seems to get worse by the day. Over the past decade I have also dragged her (and Patch the dog) to numerous windswept headlands and onto many storm lashed beaches over the length and breadth of this country (and others) chasing Mr. Marconi. So telling the story of Hall Street is much easier.

My thanks to Mike Plant for once again stepping up to read my rambling text and fix the worst of my mistakes and for Chris Neale for pitching in to help with that thankless task as well. Also Chris's newspaper research was much appreciated with the emails soon arriving faster than I could read them. But, as ever, any and all errors and omissions remain firmly mine....and finally, Mark Lloyd - another great cover. Nice one.

PHOTO CREDITS AND COPYRIGHT NOTES

refrain from enforcing the copyright which subsists in works on the grounds of public interest. For example, patent diagrams and handbooks are held to be in the public domain, and are thus not subject to copyright. If I have inadvertently tripped over an obscure copyright I apologise and will immediately update the digital image with the printers giving copyright designation.

AUTHORS NOTES - Wireless, Radio and Hertzian Waves

The term 'wireless' telegraphy, as applied to electrical systems, is actually somewhat misleading, since it implies the absence of wires in systems where wires are of course essential for both the aerial and as part of the equipment. It is generally understood that the phrase 'wireless telegraphy' describes telegraphy without 'metal connections', and because the more improved methods have in fact increasingly lessened the amount of wire used, the phrase has been allowed to stand.

Furthermore the term 'wireless communication' makes the distinction even more blurred as it does not implicitly define or imply the use of electromagnetic waves. Many early pioneers had struggled to communicate without wires used electromagnetic conduction and induction systems or indeed combinations of the two, sending signals either through water or earth. They included Loomis, Stubblefield, Steinheil, Lindsay, and Dolbear. Meanwhile Morse and Preece even went into limited short range commercial operation using systems that communicated without wires, but employed electromagnetic induction techniques. Preece described it as 'non-direct' communication. In this text the phrase 'wireless communication' or 'wireless telegraphy' unless otherwise caveated implies communication by Hertz's electromagnetic waves.

In this text the terms 'wireless' and 'radio 'mean exactly the same thing. Surprisingly it was the word 'radio' that first came into use even before Heinrich Hertz proved the existence of electromagnetic waves in 1887. Despite this very early usage 'radio' has always been regarded as the modern version of the term 'wireless'. Hence in this story, the terms 'wireless waves', 'radio waves', 'Hertzian waves', 'Marconi waves', 'aestheric waves' and even 'electromagnetic waves' all mean essentially the same thing. This story starts over one hundred and twenty years ago, so for the most part I have used the term wireless, although by the 1920s 'radio' had become the more common or accepted term. In June 1912, at the third international conference on radio held in London the term 'radiotelegraphy' formally replaced 'wireless' as the label for communication by electromagnetic waves.

I am also reminded that a young Italian called Guglielmo Marconi once called his Company..... the Marconi *Wireless* Telegraph Company Ltd, and how could he be wrong?

PHOTO LIST

Cover photographs. A montage of the Marconi Company Hall Street works, one of the ladies from the coil winding shop on the top floor of the works and one of the gentlemen from the machine shop.

Inside front cover : Guglielmo Marconi, c. 1908 *(MWT)*

1. Hall Street, c. 1899 *(MWT)*
2. Marconi plaque BT centre London
3. Marconi and equipment 1897 *(MWT)*
4. Hall Street road sign and blue plaques
5. Advert for Marconi Ignition Coils, c. 1905
6. Guglielmo Marconi c. 1897 *(MWT)*
7. George Kemp *(MWT)*
8. George Kemp with early Equipment *(MWT)*
9. George Kemp in later life displaying one of the Poldhu aerial kites *(MWT)*
10. Staff of the Marconi Wireless Signal Company Ltd, 1895 *(MWT)*
11. The Marconi engineers, attached to Royal Engineers, Boer War *(MWT)*
12. HMS *Thetis*, equipped with Hall Street wireless during Boer War
13. Early photograph c. 1903 of the original Broomfield Wireless Station
14. H.J. Round *(MWT)*
15. H.J. Round in later life, possibly at the Broomfield station *(MWT)*
16. Typical station aerial mast *(MWT)*
17. Marconi wireless equipment onboard the *Tongue* lightship *(MWT)*
18. Niton, Isle of wight wirless station interior, c. 1900 *(MWT)*
19. Frinton-on-Sea wireless school and station *(MWT)*
20. The Clifden wireless station *(MWT)*
21. The Clifden wireless station and lake *(MWT)*
22. The Clifden wireless station condenser house *(MWT)*
23. The Marconi Hall Street works in Chelmsford *(MWT)*
24. Marconi's Hall Street works, Marconi and the Hall Street engineers *(MWT)*
25. Marconi's Hall Street works, the ladies of the winding shop *(MWT)*
26. Marconi's Hall Street works, winding shop *(MWT)*
27. Marconi's Hall Street works, ground floor machine shop *(MWT)*
28. Marconi's Hall Street works, ground floor machine shop *(MWT)*
29. Marconi's Hall Street works, ground floor machine shop *(MWT)*
30. Marconi's Hall Street works, the coil drying room *(MWT)*
31. Marconi's Hall Street works, the coil mounting room *(MWT)*
32. Marconi's Hall Street works, top floor research labs *(MWT)*
33. The Dalston Works, c. 1903 *(MWT)*
34. The Dalston Works, c. 1905 *(MWT)*
35. The Dalston Works, Carpenters workshop *(MWT)*
36. The Dalston Works, Ignition Coil assembly shop *(MWT)*
37. The Dalston Works, Capacitor assembly shop *(MWT)*
38. The Dalston Works, 2013
39. RMS *Republic II*
40. Marconi wireless operator Jack Binns

41. RMS *Titanic*
42. RMS *Titanic* wireless room
43. RMS *Olympic* wireless room (MWT)
44. Mr Punch honours Mr Marconi over the *Titanic*
45. Poldhu Point wireless station (MWT)
46. Pholudu collapsed aerial array (MWT)
47. Poldhu operating bay (MWT)
48. Poldhu spark gap (MWT)
49. Lizard receiver station (MWT)
50. Marconi at the Cabot tower (MWT)
51. Wireless operating bay aboard SS *Philadelphia* (MWT)
52. Marconi magnetic receiver
53. Marconi tuning *Jigger*
54. Marconi receiver coherer
55. Horse mounted army wireless equipment (MWT)
56. Wagon based wireless equipment
57. Marconi-Bellini-Tosi radio direction finder (MWT)
58. Marconi New Street works ground breaking ceremony (MWT)
59. Marconi New Street works construction begins (MWT)
60. Marconi New Street works foundations underway (MWT)
61. The New Street build team (MWT)
62. The Completed New Street Works (MWT)
63. New Street Works visit by International Radio-Telegraphic Conference delegates (MWT)
64. New Street, c. 1920 (MWT)
65. New Street, from the air, c. 1920 (MWT)
66. New Street, main aerial mast construction team, (MWT)
67. New Street, aerial masts (MWT)
68. RMS *Lusitania*
69. RMS *Lusitania* wireless room
70. RMS *Lusitania* wireless operating cabin
71. RMS *Lusitania* lifeboats
72. RMS *Lusitania* mass graves
73. Thurso D/F station
74. Hunstanton D/F station
75. Hall Street 1899 and Hall Street 2014.
76. Hall Street works, rear of building 2014
77. Hall Street works, Alfred Cottage 2014
78. Hall Street works plaque, 2014
79. Hall Street works rear plaque, 2014
80. Hall Street works interior, ground floor, 2013
81. Hall Street works interior, ground floor, 2013
82. Hall Street works interior, top floor, 2014
83. Hall Street works interior, top floor, 2014
84. Hall Street works interior, roof detail, 2014
85. Hall Street works interior, stairwell, 2014
86. Mike Plant in front of the site of the Hall Street Wireless station 2014
87. The Back the Bid team, 2015
88. The author at Hall Street

FOREWORD

This book provides a fascinating insight into the first years of scientific and industrial activity started by Guglielmo Marconi. The early years of the Marconi company were characterised by astonishing levels of innovation and belief in technology - and that tradition of powerful, world leading innovation is traceable even today in Chelmsford, scarcely 2 miles away from the original Hall Street wireless works.

In 1940 two scientists, Randall and Boot, demonstrated in their laboratory in Birmingham University the generation of high levels of microwave power from a device named the 'Cavity Magnetron'. This was soon realised to be the critical missing component that would make radar a practical proposition for the Royal Air Force and ground forces to give a vital advantage over the German forces in the second world war. Marconi, already developing secret radar systems and producing small numbers of experimental devices at the Valve Laboratory in Great Baddow, quickly in 1942 established a Magnetron factory in farm buildings on a sleepy lane on the outskirts of the town (with the continuing farm activity providing useful camouflage). This was the origin of the company soon after named the 'English Electric Valve Company' which, through various changes became 'EEV' and then most recently since 2002 'e2v'.

e2v still maintains its headquarters and principal manufacturing site at 106 Waterhouse Lane where it employs more than 1,000 people in design and manufacturing. When I myself first walked in to 'English Electric Valve' in 1977 as a green 6th form school boy from Chelmer Valley High School curious about what went on inside a factory, I was completely blown away by the in house TV studio (for demonstrating vacuum tube television sensors called 'Leddicons'), the glowing helium neon lasers engineers were experimenting with to aid manufacturing processes, and the wide range of other mysterious – and often secret – things going on. I was hooked. And I still get that sense of wonder regularly when I stop and think of the almost impossible things the people there do.

So, from providing life-saving technology for practically all the worlds cancer radiotherapy systems, to providing the silicon image sensors that provided, in 2014, every image reported by ESA's staggeringly bold successful mission that landed on a comet, I think that spirit of innovation lives on. From the unbelievable feat in 1901 of communicating in real time 2,170 miles across the Atlantic ocean, to in 2015 the unimaginable feat of sending back the close up colour pictures of the surface of Pluto on the outer edge of our solar system , I'm confident our current Engineers still have a strong belief in technology. And over 1000 people in the town are enjoying making their living as part of that story - and making money for UK plc.

And without Guglielmo Marconi none of this would be here. We owe respect and a great debt to this astonishing individual. So from the e2v of today, and on behalf of the many 1,000's of Chelmsford people that have been part of this continuing story, I say thank you to this remarkable man.

Trevor Cross, Group Chief Technology Officer, e2v.

PREFACE

Marconi's Hall Street Wireless Works

The name of Guglielmo Marconi is associated throughout the world with the invention of wireless. He built and tried out his first wireless telegraph sets at his home in Bologna, Italy.

As he found no official support there he came to London in 1896 with his mother (who was Irish), bringing with him his equipment. He patented his system in June 1896 – the first wireless patent. He was then just 22 years old. He demonstrated his wireless system publicly, with the generous assistance of the Post Office, and a company was formed in 1897 to use his patents, called the Wireless Telegraph & Signal Co Ltd (changed to Marconi's Wireless Telegraph Co Ltd in 1901 and to the Marconi Company Ltd in 1963).

This company needed a factory and in December 1898 they leased Messrs Wenleys' furniture store (the old silk mill) to be the world's first wireless factory. Here were fashioned the little spark wireless sets for the Royal Navy from 1900 and also for merchant ships and great liners. Parts were made for the first transatlantic test using a new high-power station at Poldhu in Cornwall, to send out the signal that Marconi heard in his earphones at Signal Hill in Newfoundland on the 12th December 1901. That signal was the first step towards the commercial transatlantic wireless service of 1907. Research in those days was carried out at experimental stations at Poole Harbour and at Broomfield. Morse code (telegraphy) was used on these early services.

As income was low they diversified into motor-car ignition coils, X-ray apparatus and scientific equipment. By 1911, however, business improved greatly under the new managing director, Godfrey Isaacs, and soon this factory was overloaded and had spread across Hall Street to a plot where a 200 ft steel mast had been erected in 1909 (where the little chapel now stands). The mast was one of the Chelmsford landmarks. Double shifts were worked in the factory to make wireless gear for export to the four corners of the globe: to the Amazon basin, Thailand, South Africa and India, to both sides in the Balkan War of 1912 and even to the cable companies – Marconi's greatest competitors. All the great Atlantic liners were fitted: *Lusitania, Mauretania, Baltic, Olympic* and the ill-fated *Titanic*.

In Summer 1912 the workpeople left Hall Street for new premises in New Street. Research work continued in the huts across the street under the tall mast and war work was soon to follow. The empty building with its coat of green ivy reverted to its earlier use. Today, without the ivy, but bearing proudly a blue commemorative plaque on the east wall, the historic old building has become the property of the Essex Water Company.

Stanley Wood.
Chelmsford, 1987.

INTRODUCTION

There are only a few places in the history of the development of wireless communication that can truly claim to be important. Places where for a brief moment in time something happened there that changed the course of a technology or a science and moved it closer toward its becoming a global business. The ancient Greeks even have a word for it, *'Kairos'* - the right or opportune moment, or even the *supreme* moment.

Those who have read my previous books will know that I am fascinated by these places and these 'moments in time' where the clock ticked and history moved on. One such moment saw James Clerk Maxwell finish his treatise on Electromagnetism in 1864 and Heinrich Hertz sitting in the a dark laboratory for nearly a year staring at a tiny gap looking for an elusive spark. I think that William Preece's wireless lecture at Toynbee Hall in December 1896 was important and the world's first wireless station at Alum Bay was such a place, although sadly now lost over the Islands cliff edge. The giant stations, the like of which the world had never seen before at Poldhu Point and Clifden were also vital pieces as Marconi built the wireless age.

Then there was New Street, the world's first purpose built wireless factory. A whole new idea that led directly to the modern age of mass electronics, computer, broadcasting, television, radio, radar, mobile phones and even the Internet.

But before the giant edifice of New Street was built, a small, almost insignificant building in Chelmsford in Essex started the modern age of electronic manufacture. Within its walls, for some thirteen years the foundations were laid that would allow the Marconi Company to grow its manufacturing systems and processes. It meant that just two years after Hall Street closed the New Street works was able to rise to the challenges of a world war, training hundreds of thousands of wireless operators, wireless engineers and revolutionising the science and art of wireless communication at sea, in the trenches and with voice communication in the air.

The Hall Street works was from the outset an ambitious venture. In early 1899 the Company had no money, no orders, a growing mass of foreign competitors often backed by Governments and it had suffered continual espionage, theft of intellectual property and patent violations. To open a factory with no orders was a leap of blind faith. But part of Guglielmo Marconi's genius was taking such risks. For perhaps his greatest venture, transmitting a wireless signal across the transatlantic ocean over 2,170 miles, he relied on the new manufacturing base at Hall Street to make much of his equipment.

As his trials, tests and experiments grew the equipment that sent the first messages across the English Channel, the first wireless messages to an aircraft and the first equipment to go to war was all built at Hall Street. Much of the equipment that went into the giant Clifden station to allow Marconi to develop his transatlantic wireless service in direct competition with the undersea cable companies was built at Hall Street.

It was Marconi's Hall Street works that designed, manufactured, supplied and installed all the wireless equipment for a new generation of giant Ocean going liners - forgotten names such as the *Olympic, Britannic* and the *Lusitania* all carried Hall Street equipment and Marconi Company wireless operators. As did the RMS *Titanic*. When she set sail for her fateful first and last voyage, she carried a fully equipped Marconi wireless room and two company operators, one of whom was to lose his life. Their actions and their messages saved over 730 lives and bought about a revolution in safety at sea as every vessel over 1,600 tons had to carry wireless equipment after 1912.

By 1912 Hall Street had done its job, its experimental wireless station just across the road had also helped convince Marconi that wireless waves did not travel in straight lines and it was Hall Street that produced the first generations of Marconi's new tuneable wireless sets.

But it was now time to step out of the light and hand over to the new manufacturing giant just across town. As the giant New Street complex took over the new demands of the growing industry the Hall Street wireless station continued into the First World War, when a chance discovery and the persistence of a young Marconi engineer and other enthusiasts and radio amateurs around the country led to the development of the 'Y' service listening service whose vital intelligence intercepts of enemy wireless communication directly affected the course of the war.

Today Hall Street is a lone survivor of one of the fastest growing and most influential industries the world has ever known.

Its story is important and for a time important things happened there.

Tim Wander,
Cowes, Isle of Wight.
January 2016.

Hall Street - Timeline

December 1898 : Marconi decides to build a wireless factory at Hall Street
January 1899 : The Marconi Hall Street wireless works is opened in Chelmsford.
March 1899
Friday 2nd : Guglielmo Marconi presented his first paper to the IEE. It was simply entitled: 'Wireless Telegraphy'. The lecture was packed and over 300 people were turned away. Marconi repeated the lecture at the larger venue at Lower Exeter Hall on the Embankment on 16th March. *The Electrician* reported that Marconi's paper was 'an event absolutely unique in the annals of the Institution of Electrical Engineers'.

Sunday 11th : First call for assistance and report of an accident at sea by wireless for the sailing ship *Elbe* made from the *East Goodwin* lightship.

Monday 26th : Marconi's Wimereux wireless station on the north French coast established using equipment from Hall Street. On Tuesday 27th the first ever wireless transmission was made across the English Channel. The messages exchanged between the Wimereux and South Foreland wireless stations were the first ever international wireless messages.

April 1899
Tuesday 11th : By using the three stations at East Goodwin, South Foreland and Wimereux Marconi's experiments with 'syntonic' tuning had made sufficient progress so that South Foreland was able to communicate with East Goodwin without a single dot being received by Wimereux.

Saturday 28th : The *East Goodwin* lightship was rammed by the Steamer *RF Matthews*. Wireless was used to support an emergency at sea.

Marconi conducted trials in the English Channel with the French Navy aboard the *Ibis*, a dispatch boat lent by the French Government, carrying an aerial 70 feet high. He then joined the French Navy store ship *Vienne* and successfully communicated with the South Foreland station over 30 miles and was able to undertake his first experiments at sea with his new tuning system.

July 1899
Sunday 9th : First wireless trials for the Royal Navy. Marconi undertook the first ever commercial installation of wireless on board a Royal Navy Ship, HMS *Juno*. The equipment was all made in Chelmsford. On Friday 14th Marconi received signals aboard HMS *Juno* from Alum Bay, a world record distance of 87 miles and the first time a civilian establishment had communicated by wireless with a Royal Navy vessel.

Saturday 22nd : First use of wireless in aviation. Marconi demonstrated wireless transmission for the British Army at their School of Ballooning in Aldershot, installing a transmitter into a captive observation balloon and then using it to send signals to another smaller balloon some miles away. The information was transferred to the ground from the receiver by a wire. The tests were premature as the age of aviation was not yet born and for the next eight years Marconi's apparatus would remain land and sea bound.

August 1899
During sea trials and war games wireless proved itself to be a major asset to the Royal Navy. HMS *Juno* and HMS *Europa* maintained reliable contact over 60 miles, and could communicate up to a maximum of 74 miles.

September 1899
Tuesday 18th : Demonstration and lecture for the British Association by J.A. Fleming from a temporary station in Dover town hall linked England and France by wireless telegraphy and included a link by telegraph wire to a meeting in Italy.

Saturday 23rd : Marconi's new Hall Street works in Chelmsford received signals from the Wimereux station in France. A new world distance record between shore stations of 85 miles, of which 58 miles were overland.

Sunday 24th : Wimereux exchanged signals with the new Marconi Harwich coast station, equipped with Hall Street manufactured equipment over 83 miles. Soon after, the Wimereux station was dismantled.

October 1899
The results of the Royal Navy summer manoeuvres convinced the War Office to officially accept Marconi's wireless equipment for ship to shore and ship to ship communication.

Wednesday 11th : In South Africa, a war between the United Kingdom and the Boers of the Transvaal and Orange Free State erupted. It was to be known as the Boer War.

Monday 16th to Friday 20th : Marconi successfully reported on the America's cup yachting race in New York using wireless aboard a ship following the race.

Monday 23rd to Saturday 4th November : Marconi conducted wireless equipment trials for the U.S. Navy.

Saturday 28th : Marconi sent the world's first paid ship to shore *radiogram*; this the first ever official wireless message sent by the U.S. Navy.

November 1899
Thursday 1st : Marconi Company Hall Street manufactured wireless equipment was deployed to South Africa for the Boer War. It was the first time wireless equipment was used in a war. Marconi made plans to leave America.

Friday 2nd : Marconi onboard the USS *New York* transmits the first official naval wireless message sent from aboard a U.S. naval vessel at sea.

Friday 24th : The Marconi wireless engineering team arrived in South Africa. They were immediately conscripted into the British Army.

December 1899
Wednesday 15th : The SS *St. Paul* became the first transatlantic vessel to report its arrival by wireless and then pass on messages from its passengers. The first newspaper

to be prepared at sea from wireless news, *The Transatlantic Times,* was printed on board the SS *St. Paul*

February 1900
Sunday 18th : The first German wireless station was equipped with Marconi's wireless system at Borkum Island. It was installed by the Marconi Company direct from the Hall Street works in the Island's lighthouse. Another station was installed on the *Borkum Riff* lightship lying 18.6 miles (30 km) northwest of the Island.

Wednesday 28th : The German liner *Kaiser Wilhelm der Grosse,* carrying 1,500 passengers was the first commercial installation of wireless equipment on a merchant ship. It was fitted with untuned Marconi wireless equipment and sent its first *Marconigram* to the operating *Norddeutscher Lloyd Company* at Bremerhaven, via Borkum Island, over fifty miles away.

March 1900
Saturday 17th : After a difficult operational start the Wireless equipment in British Army service in the Boer War was transferred to the Delagoa Bay Squadron of the Royal Navy for blockade duties against the Boer forces. HMS *Thetis* became the first ship to be equipped with wireless in a theatre of war. The equipment immediately proved successful in operational naval use.

Saturday 24th : The Wireless Telegraph and Signal Company name was changed to Marconi's Wireless Telegraph Company Ltd.

April 1900
Wednesday 25th : The Marconi International Marine Company Ltd was founded.

Thursday 26th : Marconi filed patent number 7777, known as the four sevens patent, for 'syntonic' tuning.

May 1900
Tuesday 22nd : The world's first permanent wireless station at Alum Bay in the Royal Needles Hotel Station on the Isle of Wight closed down.

June 1900
Thursday 7th : Marconi's Niton wireless station on the Isle of Wight started operations.

July 1900
Monday 2nd : A contract was entered into between the Marconi Company and the Admiralty for the installation of the Marconi apparatus on twenty eight Royal Navy ships and four coast stations.

Wireless sets were sent by the Royal Navy to the China naval squadron for use during the Boxer Rising, installed on board the battleship HMS *Balfour.* This marked the first capital ship in a naval fleet to use wireless on active service.

The Marconi Company Board of Directors, after some persuasion, gave Marconi their approval and authorised funding for his huge transatlantic experiment.

October 1900 : Construction of the Cornwall transatlantic wireless station at Poldhu Point commenced.

November 1900

Friday 2nd : First wireless land station in Belgium opened at La Panne with Marconi Hall Street equipment, allowing constant contact between the station and ships sailing the route between Ostend and Dover. The mail packet ship *Princesse Clementine* stayed in communication with the shore during its entire journey.

10th November 1900 : Marconi's first wireless patent application was filed in America, but is was refused.

January 1901

Thursday 1st : The *Princesse Clementine* reported by wireless that the barge *Modora*, of Stockholm, was waterlogged on the Ratel Bank.

Tuesday 8th : Wireless telegraph experiments aboard *Princesse Clementine* were undertaken in a heavy storm, but communication was maintained the whole way from Ostend to Dover.

Saturday 19th : The *Princesse Clementine* ran ashore at Marikerke during thick fog and summoned assistance by wireless telegraphy.

Thursday 10th : The Marconi Company started operation of three wireless telegraph stations linking all the Hawaiian Islands, Oahu, Kauai, Molokai, Maui, and Hawaii except Kauai. These were the first Marconi wireless stations on American soil and all the equipment and the engineers were sent from Hall Street. The poles of the wireless stations ranged from 125 to 175 feet in height above sea level.

Wednesday 23rd : Marconi's wireless station on The Lizard in Cornwall received transmission from the Niton Station over 186 miles, a new world distance record. This provided yet more evidence for Marconi that a wireless signal across the Atlantic Ocean might be possible.

March 1901

Saturday 2nd : The Hawaiian Wireless Telegraphy Company opened for business. The rates for messages were two dollars for messages of not more than ten words and twenty cents per word for each additional word. The greatest distance over which this line operated was forty three miles from the Makena to Mahukona stations.

April 1901

Sunday 14th : During a three month trial with the French Government, the Marconi Marine Company established a successful wireless connection between Antibes in France and Calvi on the Island of Corsica, 175 km (108 miles) away.

An order was received from the Canadian Government to install Marconi apparatus at two stations on the Straits of Belle Isle.

May 1901
Wednesday 8th : HM treasury finally approved the purchase of Marconi wireless telegraphy equipment for the Royal Navy. This was the first commercial order for the Company and initially consisted of 32 Marconi wireless sets, plus the five already in service from the Boer war. The order had to be delivered and tested before the end of December 1901.

Tuesday 21st : The first British merchant Navy ship to be equipped with a commercial wireless system, the Beaver Line's SS *Lake Champlain,* set sail from Liverpool.

Tuesday 28th : Using the Antibes transmitter station, the Prince of Monaco's yacht *Princess Alice* was fitted with Marconi apparatus for a demonstration to HH Prince of Monaco, former French Empress Eugenie and many guests along with delegates of *Congres International de l'Association de la Marine.*

June 1901
Saturday 1st : Agreement entered into between Marconi Wireless Telegraph Company, Limited, and the *New York Herald* Company for the installation of Marconi apparatus at Siasconset lighthouse, Nantucket Island, and on the Nantucket Shoals lightship. The system was used for communication with ships at sea, interchange of shipping and other news between ships at sea and *New York Herald* office, and transmission and receipt of service and passenger telegrams.

Saturday 29th : A new world distance record for wireless transmission is established over 225 miles between the Poldhu Point wireless station and the newly opened station at Crookhaven, County Cork in Ireland.

July 1901
Tuesday 23rd : Agreement made with the Government of the Congo Free State for the installation of Marconi wireless telegraph stations at Banana in the Congo Free State and at Ambrizette in Angola, the Congo Free State undertaking for ten years to use no system of wireless telegraphy other than the Marconi system.

August 1901
Saturday 3rd : The first Cunard liner to be fitted with a Marconi wireless system, built at Hall Street, the RMS *Lucania* departed from Liverpool.

September 1901
Tuesday 17th : Severe gales destroyed the 'transatlantic' Poldhu wireless station antenna array. A temporary and much smaller antenna system was hurriedly constructed so the experiment could continue.

Thursday 26th : After considerable intrigue and three years after the successful tests at Rathlin Island, Lloyd's of London placed its first order with the Marconi Company to exclusively supply wireless equipment for ten of its coastal signal stations. Lloyd's agreed, for a period of fourteen years, 'not to use or permit to be used at or in connection with any of their stations any system of wireless telegraphy other than the Marconi system'.

1901 : The world's first Wireless Telegraph Training College opened at Frinton-on-Sea on the Essex Coast. Part of the syllabus involved regular tests and communication with the Hall Street wireless station.

October 1901
Saturday 26th : The *La Compagnie de Telegraphic* Company was founded, with its head office in Brussels, to develop and work with the Marconi wireless system on the Continent.

November 1901

1st November 1901 : *The Essex County Chronicle* reported that the Hall Street works was proposing to erect a second new aerial mast, 150 feet high.

Tuesday 26th : Marconi, accompanied by his two technical assistants G.S. Kemp and P.W. Paget set sail from Liverpool to Newfoundland on the SS *Sardinian*. They immediately received news that a severe gale had also destroyed the Cape Cod (South Wellfleet) wireless station's antenna array in America. Marconi's dream of two way transatlantic communication via wireless was over. Marconi shrugged off this dramatic change in fortune and sailed for the New World to attempt transmission in one direction using the temporary antenna array in Cornwall.

The Broomfield experimental wireless station in Chelmsford was operational.
December 1901 : Thursday 12th : First wireless transmission of the Morse code letter 'S' across the Atlantic Ocean by Guglielmo Marconi.

February 1902 : Marconi started conducting more organised and documented tests sailing on board the SS *Philadelphia* as she sailed west from Great Britain, recording signals sent daily from the Poldhu station showing successful reception up to 2,100 miles (3,400 km).

December 1902 : The Marconi station in Glace Bay, Nova Scotia, Canada transmitted the first signal from North America back to Great Britain.
1904 : The U.S. Patent Office reversed its earlier decision, and awarded Marconi a patent for the *'invention of radio'*.

16th March 1906 : *The Essex County Chronicle* carried an advert for 'Girls wanted for light work, not under 16 years of age to apply at the Hall Street works.'

27th July 1906 : *The Essex County Chronicle* announced that the Works would close on August 11th as the Company was moving to Dalston, where many of the employees had already started. Much of the machinery had already been removed.

8th December 1906 : *The Newsman* newspaper reported that the large factory in Dalston was now in full working order after only a few months.

15th October 1907 : The huge Clifden Station on the west coast of Ireland opened for transatlantic wireless traffic.

18th July 1908 : *The Newsman* newspaper reported that the Marconi Wireless telegraph company was returning to Hall Street, stating that the plant and machinery was already being installed and that work would restart in two weeks.

1909 : Marconi and Karl Ferdinand Braun were awarded the Nobel Prize in Physics for 'contributions to the development of wireless telegraphy'

23rd January 1909 : The RMS *Republic*, a steam-powered ocean liner, collided with the Lloyd Italiano liner SS *Florida*. She sent the first ever CQD distress call issued by wireless telegraphy, using equipment built at Hall Street, which resulted in the saving of around 1500 lives. Known as the 'Millionaires' Ship' on account of the number of well-known and immensely rich Americans who travelled by her, she was one of the largest and most luxurious liners afloat.

1910 : The Wireless Ship Act was passed by the United States Congress, requiring all ships of the United States travelling over two hundred miles off the coast and carrying over fifty passengers to be equipped with wireless equipment with a range of one hundred miles. The legislation was prompted by the *Republic* incident.

31st July 1910 : Dr Hawley Harvey Crippen an American homeopath, ear and eye specialist and medicine dispenser, was the first suspect to be captured with the aid of wireless telegraphy while crossing the Atlantic on the SS *Montrose* that was equipped with Hall Street wireless equipment. He was eventually hanged in Pentonville Prison for the murder of his wife Cora Henrietta Crippen.

16th December 1910 : *The Essex County Chronicle* reported that a new mast was being erected at Hall Street. A 180 feet high steel mast was topped with a wooden top mast to give 210 feet in total. 20 tons of cement were used for the beds of the mast and the supports.

May 1911 : The manager of Hall Street was Mr C. Mitchell, Mr H.N. Dowsett was Chief of the Technical Staff and Mr A. Eddington was the Assistant Works Manager.

26th February 1912 : Construction commenced on the Marconi New Street Works.

14th April 1912 : The RMS *Titanic* hit an Iceberg mid-Atlantic. While in distress, it contacted several other ships via wireless that saved the lives of 711 people. After this, wireless telegraphy using spark-gap transmitters quickly became universal on large ships.

22nd June 1912 : The Marconi New Street Works opened after just 17 weeks of intensive effort.

August 1912 : The Marconi Hall Street works closed

1913 : Marconi initiated duplex transatlantic wireless communication between North America and Europe for the first time, using receiver stations in Letterfrack Ireland, and Louisbourg, Nova Scotia.

1913 : The International Convention for the Safety of Life at Sea was convened that produced a treaty requiring the installation of shipboard wireless stations that have to be manned 24 hours a day.

28th July 1914 : The First World War, or the Great War began. More than 70 million military personnel, including 60 million Europeans, were mobilised in one of the largest wars in history.

The war would force the pace of technological change beyond anyone's dreams.

The Marconi Company was ready and able to meet the huge demands of manufacture, training, research and development that would to be forced upon it, due in no small part to the work of the engineers and staff at the ***Chelmsford Hall Street works.***

Advert for Marconi Ignition Coils, made at both Hall Street and Dalston. c. 1905

CHAPTER ONE

'My Name is Guglielmo Marconi'

If you were a Hansom cab driver in February 1896 you might have seen two lone figures standing on the platform of London's Victoria station; a man and a woman. They are not looking for a cab, rather they expect to be collected by a distant cousin, but they have been delayed on their journey by customs officials. A mother and son, well dressed and surrounded by luggage and travelling trunks, unremarkable apart from the fact that the young man is wearing an unusual deer stalker hat and has a thick fur coat pulled tight around his thin frame. By his feet he also has two rather incongruous black boxes.

The pair have already travelled a long way across Europe from Italy. They are tired, cold and frustrated. The hour is late and they still have a long way to go. Waiting in the shadow of the great Grosvenor Hotel, the Victoria Station yard is one of the great terminal stations for the omnibuses from all quarters of the Metropolis. This, combined with the busy horse drawn Hansom cab traffic could not have failed to impress them. Any visitor from the Continent would soon come to sense the bewildering energy of the mighty city.

The Mother's cousin, Henry Jameson Davis arrived. A first cousin, but he hasn't met the young man since he was six years old. Little did he know that his journey across London in the horse drawn Hansom cab to collect the visitors would also change his life for ever. The plan is that Henry Jameson might be able to introduce the inventor and his new ideas to the scientific community of late Victorian England.

The young man, half Italian and half Irish is Guglielmo Marconi and he has a dream of changing the world with a communication system that does not need wires. As Marconi caught his first glimpses of the unfamiliar capital, the young man, barely 21 years old, could never have dreamt how far this journey would take him.

As the new century approached, Victorian England toasted the advent of a new scientific age. Between 1885 and 1889 the German physicist Heinrich Rudolf Hertz had demonstrated that the Scottish physicist and mathematician James Clerk Maxwell's theoretical electromagnetic waves were a fact. He could generate them, receive them and even measure their properties and they quickly became known as 'Hertzian Waves' or even 'wireless waves.' Hertz was one of the best scientists of his generation, but he died in 1892 at the very young age of 36 and never saw any practical use for his laboratory experiment.

Heinrich's 'Hertzian waves' were still little more than a scientific curiosity when a number of other inventors and scientists, including Professor Oliver Lodge, started organising lectures and practical demonstrations repeating his work. But like many other early pioneers even Lodge failed to see any commercial potential in his experimental results.

Part of the problem was that the prominent scientists of the time had announced that the curvature of the earth would severely limit the range of these new waves and consequently they had no practical purpose. It was also believed that the powerful waves from any high powered long distance station would swamp the feebler waves from ordinary ship and shore stations, producing nothing but chaos in any receiver. The established scientific community was adamant that wireless communication by electromagnetic waves had no future and could never challenge the well-developed cable telegraph system and network.

In 1896 the young Italian inventor arrived on Britain's shores with a dream of transmitting messages in all weathers and at any time and place, without the use of wires. He had just two boxes filled with some barely working and rather crude wireless equipment, fresh from his experimental attic work bench, that had already been damaged by wary customs officers.

He had no staff, no finance, no offices or workshop, no scientific training, no formal education and perhaps most importantly for Victorian England, no reputation. What he did have was an enormous passion and an almost unbelievable belief in himself, his ideas and his system.

Marconi was about to embark on a determined, almost obsessive path to prove that his laboratory experiments were the basis of a reliable, long distance communication system. From the moment he first set foot in England, Marconi's public demonstrations and private tests were all designed to move his system one step further, both in technological development and in commercial understanding and acceptance.

The problem for Marconi was that the great Heinrich Hertz had shown that his invisible waves obeyed exactly the same laws of reflection and refraction as did light waves, and that in fact the only fundamental difference between the two was one of frequency. These conclusions had also been verified time and again by Marconi and other scientific workers. Yet equally beyond dispute was the fact that the data being amassed from Marconi's demonstrations and experiments at numerous wireless stations around the world showed a steady upward progress in the ranges that could be reliably achieved. For some reason Hertzian waves were breaking the 'straight line' rule. At first the difference between theory and observation was small, but it soon rose to double the predicted figure. When distances of eighty miles could be guaranteed, and Marconi could reliably communicate with ships and other stations far below the horizon this anomaly could no longer be ignored. Still the common scientific opinion of the day simply chose to doubt the authenticity of these experimental results.

At times it was only the charisma and force of character of the young Italian inventor that kept interest in the system alive. Guglielmo Marconi was convinced that the eminent men of science were wrong and that reliable long distance communication was just a matter of increasing transmitted powers and improving the sensitivity of the receivers. This belief was soon backed by his repeated observations of successful transmissions to ships lying below the horizon, suggesting that the propagation of wireless waves was far more complex than the 'straight line' or 'line of sight' theorists believed.

For the first five years after arriving on England's shores, Guglielmo Marconi had two driving ambitions that fuelled his research. He was determined to quash the persistent rumours regarding the accuracy of the ranges he claimed for his system. A practical and reliable range of at least 25 miles in all weathers was essential to his dream of providing systems to safeguard ships at sea. By the end of 1897 his system could do that and more every day and in every type of weather.

It is easy to forget that before Marconi and the widespread introduction of his reliable wireless communication system, any ship, large or small, commercial, civilian or military effectively disappeared as soon as it left sight of land. Sometimes it never returned and the owners and relatives of the passengers and crew would never know their fate.

Marconi's second goal was perhaps more ambitious. He was determined to set his system in direct competition with the long distance undersea telegraph cable companies. Until 1899 wireless communication was still largely retracing the path blazed by the cable telegraph half a century earlier, developing point to point communication, albeit now without the need for expensive connecting wires.

But developments moved rapidly. As the new century approached there were the beginnings of talk about innovations which moved beyond what the cable telegraph could do. People started to understand that the new *'wireless'* could be useful in aiding safety at sea. Marconi also saw a role for the new system in *'the* warfare of the future', and also the potential to someday compete with the telephone in providing personal communication.

By 1899 Marconi had already built the world's first permanent wireless station at Alum Bay on the Isle of Wight and new stations had been established on lightships and on the English east and south coasts.

Marconi's wireless system would soon instigate the first ship to shore message, the first shipwreck rescue, first use of the international distress signal and undertook the first transmissions to France across the English Channel. He would soon undertake extensive trials with the Royal Navy and even send his fragile experimental system to face the harsh environment and demands of the military during war for the first time in South Africa.

Throughout this period, Marconi was determined to maintain the technological integrity of his Company, never publicizing new inventions until they had been fully proved or making promises which he could not substantiate. Through it all the young Italian displayed a considerable flair for publicity and a deft hand with the press, public, military authorities and even the doubting scientific community.

Being Marconi, he tackled all these head on with a quiet reserve but ruthless dedication to the task as he moved from experiment to demonstration and from test to trials, each carefully calculated step pushing his system, himself and his team harder and harder. Young as Marconi was, his dedication and single-mindedness coupled with his gentlemanly demeanour was very different from the popular Victorian image of the 'mad inventor', and completely different from some of his more illustrious competitors. His continuous successes despite the many obstacles, inspired loyalty in his small workforce of engineers, most of whom had learned their trade in the established Victorian business of telegraph cables. Guglielmo Marconi once said, with some modesty:

'I have striven to give the world improved and cheaper means of communication by means of electrical transmission through space.'

Guglielmo Marconi c. 1897

Marconi patented his system in June 1896 with the first ever true wireless patent when he was just 22 years old. His Company, formed in 1897 to use his patents, was called the Wireless Telegraph and Signal Co Ltd. This was changed to Marconi's Wireless Telegraph Co Ltd. (often known by its initials MWT Co) in 1901.

The Company was set to develop the science of wireless communication. Of course such advances would needed talented men of vision and passion who were prepared and able to take on and meet the challenge. Marconi built around him a team that included some of the best engineers in the world, drawing people from many disciplines to work at his side as he built his Company.

But out of all the early pioneers and scientists it was Marconi alone who had the personal drive and single minded belief to develop reliable wireless communication against enormous odds.

CHAPTER TWO

MARCONI'S HALL STREET WORKS

The World's First Wireless Factory

By the end of 1898 Guglielmo Marconi has spent two hectic years testing and demonstrating his system for wireless communication across the world. He had run countless experiments and trials for the Royal Navy, British Army, United States Navy (and Army), Trinity House, GPO, Parliament, the British Royal family, scientists and newspapers.
He believed that he had proved his case. His wireless telegraphy system was no longer a laboratory experiment. Marconi had developed a practical communication system that did not need wires.

As the turn of the century drew near, to the outside world it looked as if the advance of wireless communication was unstoppable. Marconi was internationally famous and each day seemed to bring fresh news of another successful trial or world record using his communication system.

To meet the *anticipated* demand for new equipment the Marconi Company had started to urgently seek new premises for manufacturing and administration. The Company's existing Head Office at 28 Mark Lane in the City of London was already overcrowded and could never support the proposed expansion or any form of large scale manufacturing.

The world's first wireless station, built by Marconi at Alum Bay on the Isle of Wight was still based in the rented ground floor rooms of the Royal Needles Hotel. Marconi had established an experimental and research base at the Haven Hotel near Poole but again in rented premises within a public hotel.

The Company had now taken on the task of building and commissioning a series of coastal wireless stations while it was still operating and staffing these two stations. At the start of 1899 the Haven Hotel station had only been in operation for two months and its limited space was earmarked solely for research and development. But the construction and maintenance of all these stations, together with the considerable sums for wages, travel, materials and equipment being spent by Marconi and his engineers were rapidly draining the Company's coffers. Marconi had tried to approach the American market but had met considerable resistance.

Despite the advantages of considerable favourable publicity, world-wide fame and the fact that the Company could now offer a viable, if limited, wireless telegraphy communication system, the Company's receipts were insufficient to offset costs and cash flow was becoming critical. Marconi's new Company was close to financial collapse. The Company needed a substantial order, but also realised that when it came they would be ill-equipped to respond to it.

As the British Admiralty continued to evaluate the new Marconi wireless sets it became clear that the Royal Navy was very much setting the agenda for the development of wireless communication in Britain. It was designing a training syllabus; it was pushing the technology to match its needs of maritime communication, especially with regard to ship-to-ship communications; it was setting a policy in relation to wireless developments, and last and most definitely not least, it was providing a market, possibly the sole market, for wireless communications and the Marconi Company during this embryonic period in wireless history.

The connection between the Admiralty and the Marconi Company dated back to 1896 when Marconi first gave his early demonstrations to officers from the Royal Navy including Captain Henry Jackson. Jackson had been experimenting with wireless telegraphy himself and was probably the first person to actually signal from ship-to-ship using wireless telegraphy. But as a serving career officer Jackson had no interest in commercially developing or patenting wireless, being solely interested in ensuring that the British Navy had the best system to allow it to effectively defend the nation. He had continually advised Marconi on how to adapt his wireless system to make it more suitable and reliable for maritime use and supported the integration of Marconi's wireless system into the Royal Navy, in parallel with the development of his own system.

Marconi was convinced that the Royal Navy would place a substantial order for wireless equipment and once they did, other major organisation including Trinity House would follow. He also knew that as part of any order he would need to be able to demonstrate to any potential client that he had the capacity to build their equipment to tight timescales.

As usual Marconi made the decision to look for manufacturing premises very rapidly, his choice being a large building in the town of Chelmsford in Essex. The exact reason why Marconi chose Chelmsford is unclear. From his earliest research work in London Marconi had found that his experiments were often plagued with electrical noise from tramways and lifts and Chelmsford was still reasonably free from such problems. Chelmsford had other advantages as buildings were far cheaper outside London and the county of Essex is very flat, ideal for wireless experiments and erecting aerials. Chelmsford also had a direct rail link into the capital and was reasonably near the Port of London whose huge volume of shipping represented one of the Marconi Company's immediate potential market places. Also Marconi did not particularly like London. Other than the social life and academic institutions the combination of smoke, smog and fog from millions of coal fires, along with the unpleasant odour of the streets, horses and unwashed bodies meant that Marconi preferred the open spaces of Essex. In the end the choice was probably influenced by a combination of all these factors, or perhaps it was just that suitable premises became available at the right time.

It is also possible in 1896 when he first arrived in England that Marconi had travelled to Chelmsford to seek advice from Colonel Crompton whose electrical engineering company had been founded in Chelmsford in 1878. He may even have met another Chelmsford based electrical engineering company, Christy Brothers. Crompton's Electrical Engineering Works were amongst the best in the country and the Colonel and James Christy were good friends in the Chelmsford Industrial community. It also meant that Chelmsford had a large skilled workforce already familiar with electrical systems manufacture and an established power supply network.

Marconi's choice for his first wireless equipment manufacturing premises was a large building located in Hall Street. When it was built by John Hall in 1858 it was a state of the art steam driven silk mill. John Hall had worked the mill until 1861 when the Government repealed the silk duties and French imports soon ruined the Essex silk industry. The Hall Street mill closed in 1863, but Samuel Courtaulds of Braintree, who survived the disaster, ran the mill from 1865 until 1892. It then became Messrs. Wenleys' furniture storage depot. The site was ideal for Marconi's requirements especially as the engine house still existed as did the line shafts to power equipment inside the works. The site also had a house attached now called Alfred Cottage which provided both accommodation and additional office space. (This building is sometimes referred to as having been named after Marconi's brother, Alfredo Marconi, but it is shown on maps before 1898 and probably dates back to the managers house of the original silk mill.)

In January 1899 Hall Street became the world's first wireless equipment factory when Marconi took out a 20 year lease. At first the site employed just 20 men and 2 boys. The factory was set up to manufacture spark wireless sets and coherer receivers to Marconi latest designs, but wireless was still in its infancy. Initially the Company struggled for income, so the factory had to diversify into manufacturing motor-car ignition coils, X-ray apparatus and other scientific equipment in an attempt to balance the books.

But until the Hall Street works came into operation, any wireless equipment built had been constructed by hand as required, using various modified apparatus bought from established scientific laboratory suppliers. Marconi and his assistants hand constructed other specialist parts but they could never hope to cope with quantity production of commercial equipment.

Marconi knew that to fulfil any commercial order, especially for the Royal Navy, all equipment parts would have to be interchangeable and all apparatus had to be built to a high quality and designed to be easily serviced and maintained.

It was a new industry for a new world.

In reality the Marconi Company was still a small organisation, developing equipment in an unknown field for customers yet to be won. During these early years any new engineer considered himself to be part of an elite group, learning the new science of wireless communication as it was invented, perhaps even working alongside the young Guglielmo Marconi himself.

The formation of the Marconi Hall Street manufacturing works now put the Company's whole manufacturing system on a much more formal level, with new departments responsible solely for their own specific areas of research, design and manufacture. The condenser and winding shop, mounting and machining shops all found their way into the new factory moulded into an organised commercial concern under the personal supervision of the new Works Manager, Mr. E.T. Priddle.

The Newsman newspaper on 14th January 1899 reported:

GOOD NEWS FOR CHELMSFORD

A WIRELESS TELEGRAPHY FACTORY TO BE ESTABLISHED

'Messrs. Wenley and Son, furnishers and up-holsterers, have let upon lease their factory in Mildmay Road, [this road adjoins Hall St. and is sometimes used as the address for the Marconi factory] which formerly belonged to Messrs. G. Courtauld and Sons, to the Wireless Telegraphy and Signal Company for the purpose of their manufacturers. The new system of wireless telegraphy, as doubtless many of our readers know, was invented by an Italian scientist named Marconi, and the managing director of the company is Mr. Henry J. Davis. It is believed that to begin with the company will employ 20 hands. Experiments with the company's system whether in telegraphy or signalling, are stated to have given entire satisfaction, among others, to the Postal authorities and to Lloyds; and their instruments were used on the Prince of Wales's yacht to communicate with Osborne House when his Royal Highness lay in the Solent suffering from a fractured kneecap. It is considered that the invention will be of great use to the public, especially in connecting lighthouses and lightships. Arrangements are being made under the direction of Signor Marconi at the South Foreland lighthouse and on board the South Goodwin lightship for a series of experiments in wireless telegraphy. If the experiments prove satisfactory, it is stated that the wireless system will be adopted forthwith as a means of communication between the South Foreland lighthouse and the South Sands Head lightship. The points of communication are about three miles apart. We understand that the first will take possession of the premises in Mildmay Road, Chelmsford, at the end of this month.'

One of the key members of Marconi's team who was tasked with the commissioning of the new Hall Street works was Marconi's right-hand-man, George Kemp. George Stephen Kemp was born in Kent in 1857. He served as an electrician and instructor with the Royal Navy before working for the Post Office and its Chief Engineer, William Preece. When Preece realised the great potential of Marconi's system he decided to throw the weight of the Post Offices resources behind him and George Kemp was one of the first men he put to work alongside the young Italian inventor.

Kemp first met Marconi in July 1896 during the early demonstrations on the roof of the Post Office in St. Martin's Le Grand. When Marconi looked over the ornate stone balustrade, he saw a short man, (Kemp was only five feet tall) with thick curly red hair and a handlebar moustache watching him curiously from the pavement below. He caught Marconi's eye and shouted up, 'What are you doing there?' Marconi called back, 'Come on up and I'll show you.' The onlooker arrived on the roof with such remarkable speed that Marconi believed he had scrambled up the eight storey drainpipe. The moment George Kemp reached the rooftop he knew that he would work for Marconi. He eventually 'signed on' with the fledgling Marconi Company as Marconi's 'first assistant' in November 1897. From that moment Kemp devoted himself entirely to wireless telegraphy, becoming Marconi's inseparable *first assistant*, friend and confidant for the next thirty six years. Kemp was at Marconi's side for his most memorable achievements, including the first wireless transmission across the Atlantic Ocean, and worked tirelessly on numerous windswept headlands and at sea, often under the harshest of conditions, erecting aerial masts and installing wireless equipment. More than anyone it was probably Kemp who physically built the wireless age.

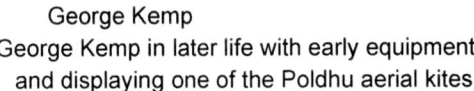

George Kemp
George Kemp in later life with early equipment
and displaying one of the Poldhu aerial kites

The *Essex County Standard* newspaper carried a in-depth interview with George Kemp about the new Hall Street Works on 29th April 1899.

WIRELESS TELEGRAPHY IN ESSEX
SIGNOR MARCONI IN CHELMSFORD

'It was mentioned some weeks ago, in the *Essex County Standard*, that Chelmsford was likely to be the head-quarters of the new 'wireless telegraphy' works in England. This important and satisfactory news is duly confirmed. A huge pole is to be erected, in a week or so, at the corner of Hall Street and Mildmay Road, Chelmsford, for the purpose of experiments. The pole is to be 150 feet in height, and a series of experiments will be made in the new means of communications. It is believed that the experiments will be in the nature of messages transmitted to the coast, probably to Harwich.

Signor Marconi paid a flying visit to Chelmsford on Wednesday to inspect the arrangements made for the establishment of his headquarters in the town. He has acquired the large building at the corner of Hall Street and Mildmay Road, which has been used until recently by Messrs. Wenley: and here in future the delicate instruments used in connection with the system will be manufactured. Operations have been commenced, though only a small staff of workmen are at present employed.

INTERVIEW WITH MR. KEMP

Mr. Kemp, who has been assisting Signor Marconi, in his experiments, is in charge at Chelmsford and has been interviewed this week by a representative of the '*Essex Weekly News*' to whom he was very communicative, being evidently in high spirits over the success of the recent cross-Channel experiments.

'Yes'; said Mr. Kemp in reply to a question, 'you see we have made a start. This is going to be our works, and it will also be an experimental station. We have been buying all the instruments we require from different firms, but in future we shall make them here ourselves.

'Then Chelmsford will be your head-quarters?'
'Well, yes; although we have two important experimental stations elsewhere – at Alum Bay and Poole Harbour.'
'The objection to wireless telegraphy so far,' I ventured to remark, 'has been this: That anyone properly provided with instruments can intercept the message and learn what it is all about.'
'That has been the case,' replied Mr. Kemp. 'But last Sunday I assisted in experiments which proved that a message can be concentrated on a given point which it is intended for. I was at Boulogne, at the station there. A vessel engaged in the experiments was in mid-Channel, and another operator was at the South Foreland. Mr. Marconi sent messages to me from the vessel which were not received at the Foreland; messages were sent from the Foreland to the vessel, which I did not receive; and I sent messages from Boulogne to the Foreland, which those on the ship could not decipher.'
'How is it done?'
It is simply done by tuning – two instruments not responding to each other unless properly tuned'.
'Does Signor Marconi hope to revolutionise telegraphic communication as it has existed by wireless telegraphy?'
'Oh no', Mr. Kemp promptly replied.
'You cannot have anything better than a good telephone, for instance, for purposes of communication.'
'But wireless telegraphy will be simply invaluable where ordinary communication by telegraph or telephone cannot take place – from lighthouse to the shore, between ships at sea, and on land where a wire cannot be laid.'
'I understand you are going to erect a high pole here?'
'Yes: we intend putting one up 150ft. in height, so that we can signal from Chelmsford to the coast. I have been down to Maldon today, to a yacht builder, in order to get a mast that will answer to our requirements. The other station will be somewhere in Essex – probably at Harwich.'

In reply to a request for information on the subject of 'Wireless Telegraphy' Mr.

Kemp stated that 'Signor (as he calls the inventor) 'Mr'. Marconi's great patent is the vertical wire. Messages have been exchanged by means of 'reflectors' but only for short distances. But by means of the vertical wire they can be transmitted for long distances – for example, between England and France – and the higher the wire is carried the longer the distance the message can be sent.

Now this is what happens. As you know, sound travels in successive airwaves. What we do, is to erect a high pole, crossed at the top by a short spar. One end of the wire – the vertical wire – is fixed to the spar, and the other end is connected with the receiving and transmitting instruments below. When the operator transmits a message he taps the instrument, and each tap causes the wire, which is highly charged with electric currents, to set up a wave. This electric spark, or wave shock, is instantaneously recorded by the receiving instruments at the other station wherever it may be. The message is sent by means of long and short taps, or long and short waves, which answer the same purpose as dots and dashes in ordinary telegraphy. But when the system is more developed, we hope to be able to send letters.'

'How long has Signor Marconi been engaged in these experiments in England?' I asked. 'Three years, next June.' Mr. Kemp answered.
'He first sent messages by means of two reflectors, erected on top of the General Post Office in London. But since then he has invented and patented the vertical wire.'
'And you have assisted him all along?'
'Yes', Mr. Kemp replied with a touch of pride.
'I have been engaged in the experiments from the outset.'
'And you think there is plenty of scope for the future development of the system?'

'Undoubtedly. We are only yet in the experimental stage. You must come when we have the big pole erected and the instruments at work.

We will show you some things that will astonish you.'

Staff of the Wireless Telegraph and Signal Company, 1895

From the earliest days of his experiments Marconi built a team of dedicated technicians who reported directly to him. There were twenty in 1900 and by 1906 that number had increased to thirty two. All were young men, more or less the same age as Marconi, most had studied electrical engineering or had experience in the telegraphic sector or in power plants. Many of them went on to hold important positions within the Company, whereas Kemp who was 17 years older than Marconi, kept his role as Marconi's irreplaceable right-hand man until he died in Southampton in 1933. Kemp was very efficient at solving all types of practical problems, kept a detailed work diary and was even in charge of Marconi's personal diet.

In September 1899 a wireless transmitting station was established on the other side of Hall Street to test equipment as it came off the production line and the tall Hall Street mast soon became one of Chelmsford's landmarks. The Hall Street wireless station, along with a new station constructed at Dovercourt near Harwich, brought the total number of wireless stations built by Marconi from December 1897 to May 1900 to twelve. These included Niton (call sign NI), Haven Hotel (HH), Lizard (LD), Poldhu (CC), Chelmsford (CD), Holyhead (HD), Caister-on-Sea (CS), North Foreland (NF), Withernsea (WS), Rosslare (Ireland RL), Port Stewart (from 1902 Malin Head, Ireland MH) and Crookhaven (from 1902 moved to Brow Head CK). All the equipment for these wireless stations was built at Hall Street.

The Newsman newspaper on 13th July 1899 reported:

CHELMSFORD THE CENTRE OF WIRELESS TELEGRAPHY

'The selection of Chelmsford as the head-quarters for carrying out experiments connected with Mr. Marconi's system of wireless telegraphy promises to be a fortunate thing for the town. In a chat with one of our representatives, Mr Priddle, the courteous manager of the works, explained that it is expected to employ some 300 hands. It is quite within the bounds of possibility that Mr. Marconi himself and Mr. Davis, the managing director of the concern, may live at the county town; while it is practically certain that Mr. Kemp, who has been associated with Mr. Marconi throughout the experiments, and Dr. Murray, Mr Marconi's head assistant, will reside either in or near the borough.

Messrs. Wenley and Son's premises in Mildmay Road, which have been acquired, are now being converted into a factory by the introduction of lathes and machinery requisite for manufacturing the instruments used in wireless telegraphy. A pole or more accurately speaking, a ship's mast – 150 feet high is shortly to be erected in the field at the rear of Messrs. Gray and Son's malting, and from this communication will be established with the east coast, probably Harwich.

Of course the idea of wireless telegraphy is not a new one, but Mr. Marconi has during the past three years perfected it and converted it into a practical and, no doubt, commercial success. The message, it is said, is sent by means of electrical waves which pass through the air. At first it was feared that these might be intercepted, but Mr. Marconi has succeeded in preventing this by attuning to one another the two wires, which are suspended obliquely from the top of masts to the transmitter and receiver at the points of communication. Those who play the pianoforte will have noticed that upon striking a note a responsible vibration will often be heard upon the wire by which some picture in the room is suspended. The principle in the case of wireless telegraphy is somewhat similar. The operators will know the 'note' of different instruments, so to speak, and set their instrument accordingly.

Extraordinary as it may seem, intervening obstacles appear to be no barrier to the proper transmission of the message, which, for example, can be sent from one room to another with doors closed, or from one town to another even though hills, buildings and other projections are in the line or route. The code used is the Morse or dot and dash, and messages can be dispatched and received at one and the same time.

The further developments of the system will be watched with interest all over the world, but nowhere with greater interest than at Chelmsford, which owes wholly and solely to the excellent railway facilities provided by the Great Eastern Railway its selection as a centre for such an important undertaking.'

In the first few months of operation the Hall Street works had also provided the equipment for Marconi's experimental station on the French coast at Wimereux with which he had successfully crossed the English channel on Tuesday 27th March 1899. The messages exchanged between the Wimereux and South Foreland wireless stations were the first ever international wireless messages. On 23rd September 1899 the French stations signals were also received at the newly built Hall Street works wireless station in Chelmsford, eighty five miles away, of which fifty eight were over land; a world distance record for wireless signals at the time.

Through all Marconi's trials and demonstrations the Royal Navy had always been an interested observer, but never an active supporter and they had yet to show any real financial interest in the new system. Captain Henry Jackson and Marconi had become friends and corresponded regularly, but Marconi knew that Jackson's foremost priority would always be what was best for the Royal Navy.

As the New Year of 1899 approached, all technical objections to the use of the Marconi system in warships had been answered. But it was still not a foregone conclusion that Marconi's system would be chosen, as the Royal Navy still had huge concerns over costs, commercial terms and patent issues. There were also concerns being raised about Marconi's close relationship with the Italian Navy (to whom Hall Street had already supplied four sets) and more recently the French Navy.

But after successful trials with Trinity House and the cross-channel demonstration, formally announced in a letter to the Admiralty in March 1899 (coinciding with Marconi's successful IEE paper), by July 1899 the Admiralty became ever more anxious to assess Marconi's latest wireless telegraphy equipment for possible operational use. Marconi's increasing reliability and progress with tuning began to make it obvious that wireless would be useful, if not vital, for warships operating at sea.

The Navy was eager to start sea trials and Jackson was instructed in an Admiralty Letter dated 1st July 1899 to arrange for two sets of Marconi apparatus to be supplied and fitted in ships for the annual Summer Manoeuvres. Fortunately Captain Jackson's tour of duty at the British Embassy in Paris was coming to a close. As Jackson had established a good rapport with Marconi and was familiar with his apparatus and wireless telegraphy in general, he appeared to be the obvious choice to take charge of the trials. His service as Naval Attaché officially ended on 10th July. He was appointed to HMS *Juno* for the manoeuvres on 11th July 1899.

Even though the proposed navy trials represented a considerable investment in manpower and costs for the Company, it jumped at the chance to give the Admiralty two sets of apparatus. The Company was desperate to seek its first lucrative Governmental market for its products. Marconi knew that these potential markets, that embraced both mercantile and naval fleets world wide, offered vast and positive publicity, which, ideally boosted by an order from the Royal Navy, would prove vital for future sales.

Marconi agreed to lend the Admiralty the apparatus it needed for the manoeuvres without cost, other than that of carriage for the sets and the expenses of the Company's technicians who would install and maintain the equipment. Marconi also agreed to defer the general question of payment terms for any equipment the Navy might wish to purchase until a later date. He even shut down the important syntony (tuning) experiments at the French Wimereux station so that he could personally lead the team. On 9th July Marconi boarded the cruiser HMS *Juno*, part of 'B' Fleet, which lay at Devonport dockyard, and installed the first ever commercial wireless equipment on a naval vessel. All the sets were built at Hall Street A second set was also put aboard ready for installation on the battleship HMS *Alexandra*, the flagship of Admiral Sir Compton Domvile, for installation when the two vessels later anchored together in Torbay on 16th July.

On 14th July 1899, Marconi reported that he was picking up the first signals ever received by a Royal Navy ship from a land station. These were emanating from the Alum Bay Needles Hotel station 87 miles away, another new world distance record. As a direct result of these successful transmissions, the Admiralty immediately requested that a third wireless system be installed on the cruiser HMS *Europa* before the August manoeuvres commenced. Marconi sent two of his assistants on board with the necessary instruments, which included a 10 inch coil, two specially screened relays, tappers, a Morse code tape inker and cells. A battery of 98 Siemens dry cells, arranged 14 in series and seven abreast was used with the coil. One relay was adjusted to work up to 15 miles, the other, more sensitive type for greater distances. They were adjusted by Marconi himself, and remained in perfect adjustment during the entire manoeuvres. Marconi installed his latest tuneable system (known as the Marconi *jigger*) developed at the Haven Hotel; the induction coil was thickly coated with paraffin wax, and two separate good 'earths' were used, one for the coil and one for the coherer.

The extended war games in the English Channel began, with wireless equipment installed in the battleship *Alexandra* and the cruisers *Juno* and *Europa* for the entire manoeuvres. The Europa led a squadron of seven cruisers and the 'B' fleet was duly dispatched to hunt for a convoy, the enemy 'A' fleet, at a given rendezvous. With Juno acting as both a scout and a wireless link with the other vessels and the shore, the Admiral's flagship HMS *Alexandra* remained with the slow moving battle fleet.

With Captain Henry Jackson in command and Marconi himself aboard the flagship HMS *Alexandra,* the wireless systems allowed the *Juno* and *Europa* to maintain regular contact despite the poor sea conditions. Continual signalling in all weathers by night and day saved many hours of fruitless steaming for the entire squadron and strategic information exchanged by wireless communication gave that section of the fleet a tactical advantage of over three hours. During the August manoeuvres the *Europa* and *Juno* reliably maintained communication over 60 nautical miles with the Juno and Alexandra working over 45 miles. The maximum distance achieved at sea was 74 miles, but on one occasion a fleet signal was relayed through the *Juno* to the *Europa* over a distance of 105 miles by using one ship as a relay point.

The trials were a huge success. Captain Jackson emphasised that the Royal Navy had much to gain and nothing to lose by adopting the system; if the Navy did not adopt it and other navies did, the Navy's numerical superiority would be effectively reduced. Unlike many of the other technological challenges which confronted the Royal Navy, wireless telegraphy offered increased efficiency and effectiveness at comparatively little cost.

The Royal Navy 'peace' manoeuvres of 1899 were to prove a turning point in Marconi's career. Wireless Telegraphy was not the decisive factor in the 1899 Summer Manoeuvres, but its seaworthiness, serviceability and strategic value were all successfully demonstrated at sea aboard operational warships. Of all the tests and experiments to date, working with the greatest Navy in the world, at sea, with Royal Navy personnel in operational battleships not only completely altered his personal relationship with the Royal Navy but gave the young inventor an unshakeable confidence that he was on the right course. The manoeuvres were also the decisive event which led the Admiralty to adopt wireless telegraphy, though battles with other departments would delay its full implementation for two more years.

During the sea trials the reliable signals sent from HMS *Europa* to HMS *Juno*, achieved when both were underway and sixty or more miles apart, profoundly focused Marconi's mind. Alum Bay had also signalled *Juno* over 87 miles. Until that time Marconi was still convinced from all his experiments at Alum Bay, Bristol and especially his last trip to Salisbury Plain that the transmission range for his system was mathematically related to the height of the aerial. This has even become known as 'Marconi's Law'. The Royal Navy trials wrecked all of his various formulae. What could have been considered a freak result during his tests at La Spezia in Italy, where he transmitted to an Italian Navy ship that lay well below the horizon, was now confirmed; for some reason wireless didn't travel in straight lines.

During the first half of 1899 Marconi had increased the reliable range of the wireless apparatus on a ship while underway from 18 to 72 miles, and boosted the speed of transmission to twenty words a minute. The channel had been crossed with ease and the resultant publicity had been superb. Marconi's results at sea during the Royal Navy manoeuvres convinced him that it was time to approach the American market. The lack of any orders and the ongoing battles within the Admiralty and the Post Office made it imperative.

During Marconi's successful reporting of the Kingstown sailing regatta in 1898 a representative of the New York Herald, Milton V. Snyder had been in Ireland to witness the event. Snyder reported what he had seen and heard to the owner of the *New York Herald*, James Gordon Bennett, himself an enthusiastic yachtsman. He told him how the *Dublin Daily Express* had posted the wireless bulletins in the window and how the new wireless system had transformed the reporting of the race. At a subsequent meeting with Marconi and Jameson-Davis in London an agreement had been forged that if the English Channel and Royal Navy trials were successful and his system range improved then Marconi would agree to come to America to repeat the Irish yacht race experiment for the America's Cup race in 1899, sponsored by the *Herald.*

Marconi sent a telegram to the newspaper on 12th September 1899 and the *New York Herald* attracted international attention when it published Marconi's acceptance to report the 1899 America's Cup Race, between the yachts *Shamrock 1* and the *Columbia* by wireless. Marconi was going to America for the first time. He relished both the challenge and the trip as he loved sea travel. The forthcoming trip held great business promise as well. The Company planned sea trials with the U.S. Navy and contemplated forming an American company on the back of the publicity that would be generated by the America's Cup coverage, which would be far greater than the Kingston Regatta.

The trip was expensive, but Jameson-Davis decided that the potential rewards were worth it. On the ten day journey to America Marconi and his Directors were accompanied by Marconi's assistants, Charles E. Rickard, William W. Bradfield and William Densham, all skilled operators and expert technicians. They sailed on 11th September from Liverpool on board the Cunard liner *Aurania*, which arrived in New York on the 21st. Marconi arrived to a wild reception, although initially many did not recognise the young Italian as he queued to leave the liner with the other passengers; he had grown a moustache since his time in France, possibly in an attempt to add a little more gravitas and even age to his appearance. On docking Marconi was obliged to answer hundreds of questions from the reporters who crowded around him at the dock and later at his hotel. Marconi had prepared a statement. Slightly overawed by the pressing crowd, Marconi seemed half-confident and half-defiant. As he came down the gangplank he told the waiting press:

> 'We will be able to send the details of the yacht racing to New York as accurately and as quickly almost as if you could telephone them. The distance involved is nothing, nor will hills interfere.'

Marconi and his colleagues checked into the elegant Hoffman House Hotel on Broadway and 24th Street in Manhattan. They had just begun unpacking when the hotel's steam boiler in the basement exploded and a frightened guest claimed that it was Marconi and his mysterious equipment from Europe that had caused the problem.

To allay the hotel's fears the Marconi team opened their trunks, carefully packed at the Hall Street works, to reveal the inert wireless apparatus within, but they soon realised that the most important trunk was missing. It contained the coherers and other essential parts for the trials and without it Marconi would be forced to cancel his coverage of the America's Cup. Hurried searches by custom officials proved futile and an uncharacteristically temperamental Marconi declared that he would now return to England on the next ship out of New York. His team calmed him. Bradfield and another assistant raced back to the wharf by horse-drawn cab to try to locate the trunk, but failed. They returned to the hotel.

Luckily Bradfield recalled that another Cunard liner had sailed from Liverpool for Boston on the same day that the Aurania left. He had a hunch that the missing trunk might be on that boat. Robert E. Livingston, a Herald reporter, volunteered to travel to Boston by train to search both ship and dock. Bradfield was right, a day later a telegram arrived at the hotel, the trunk was in Boston. With order and his humour restored, Marconi spent the next few days sightseeing in New York. After spending much time at the Custom House, he went to the top of the St. Paul building to get a bird's-eye view of New York's 'monster 'buildings; he was impressed with the swift moving lifts, and amused that his American hosts called them elevators. As he looked about he said, 'I'm not frightened that your big steel buildings will stop wireless.'

While he was in America his trip, or more correctly the success of his system had attracted considerable interest from the United States Navy. At first they had attempted to purchase equipment but the Marconi Company's policy was always to lease rather than sell its equipment. The Company's delicate finances also dictated that the Navy would have to pay for trials. In the event the United States Navy decided it would be cheaper to simply 'oversee' Marconi's as he reported on the race, effectively gate crashing the America's Cup demonstration.

Following Marconi's arrival in New York on 21st September, arrangements were rapidly made for a group of U.S. Naval officers to 'witness' the operation of his equipment during the yacht races. The Navy Department designated a group of observers, soon known as the 'Marconi Board', which consisted of Lieutenant Commanders J.T. Newton and E.F. Qualtrough, and Lieutenants John B. Blish and G.W. Denfeld, USN. All four were electrical experts and were well qualified to pass judgment on the young Italian and his equipment. In order that each officer could thoroughly investigate the operation of the system, the instructions required them to exchange posts and each to submit his own independent observations on the nature and operation of the Marconi equipment. An ever-alert U.S. Army Signal Corps also decided to join the demonstration. Special Orders No. 213, Headquarters of the Army, dated 12th September 1899, directed Sergeant Walter R. Taylor to temporary duty at the Highlands of Navesink during the yacht races.

To demonstrate his system for the waiting U.S. Navy observers Marconi set up his equipment on the SS *Ponce* and several other vessels. From the *Ponce* Marconi would follow the racing yachts. Shore stations were located in the lighthouse at Navesink, N.J., and in a building in Manhattan. While the light keepers and signalmen silently watched, the technicians raised a mast for an aerial. They installed a receiver in the lighthouse next to its normal telegraph equipment.

On 28th September, the following message from Lt. John Blish, one of the observers assigned by the Navy to report on the Marconi demonstrations, was sent from the *Ponce* to the Navesink shore station. Transmitted personally by Marconi, it is considered to be the first official U.S. Naval radio message, and was also the first paid ship-to-shore radiogram:

> 'Bureau (of) Equipment, Washington D.C.
> Steamship *Ponce* under way in naval parade, via Navesink Light Station.
> Mr. Marconi succeeded in opening Wireless Telegraphic communication
> with the shore at 12:34 P.M. Experiments were a complete success.
> Blish, Lieutenant, USN.'

This message was typed on board the SS *Ponce*, on the ship's stationery to Bureau Equipment, Washington by Lieutenant Blish of the United States Navy. It survives to this day as after being handed to Marconi for transmission, it was placed in Marconi's personal files. Lieutenant Blish's reported:

> 'This is no experiment; Signor Marconi shows by his every move that
> he has passed the stage of uncertainty. He knows exactly what to do and
> just what the results will be. What struck me most forcibly was that the
> system was prompt, certain and satisfactory from the beginning.'

As the Navy parade ended Marconi returned to the prime purpose of his visit to America, reporting the America's Cup yacht race. He took personal charge of the trials and supervised the reporting of the races using his equipment to provide up-to-the-minute information.

One set of receiver equipment was put aboard the cable ship (CS) *Mackay-Bennett* and was operated by Thomas Bowden, assistant telegraphist to Marconi. The cable ship was moored off New York City right over the New York-London transatlantic submarine cable. At the time of the race, the plan was for the *MacKay Bennett* to raise the cable from the bay bottom, splice into the line, and then with a telegraph key immediately relay Marconi's transmitted narrative to London and Paris. Another set was placed aboard the Puerto Rico Line vessel the SS *Ponce*, but part way through the series of races it had to be moved to the SS *Grande Duchesse*. These ships acted as the sea terminals.

From the deck of the observation ship which followed the yachts, Marconi transmitted the signals to Bowden who in turn sent them to another Marconi engineer, W.W. Bradfield who was in the *New York Herald* building. From here the reports were transmitted over the land telegraph and via Commercial's cables to the UK. As each update reached the newsroom, the editor's awe intensified. Never in history had an event been tracked or reported in this manner. Over 4,000 words were sent and received in less than five hours between the ship carrying the apparatus and the shore station. The messages were then transmitted over land wires to the papers. The next issue of the *New York Herald* proclaimed: 'Marconi's Wireless Telegraph Triumphs.'

The *Ponce* was reported as having transmitted approximately 2,500 words during the first day, at an average speed of about 15 words per minute. One speed test produced

31 words in one minute and 50 seconds, or about 17 words per minute. It was estimated that the extreme distance at which messages were sent and received was about 17 miles. Another estimate stated 17 miles with a 120 foot aerial and 24 miles with a 150 foot one. During the course of the races it was claimed that 1,200 messages containing an approximate total of 33,000 words were transmitted and received.

Overall the race course was fifteen miles from the Sandy Hook lightship to the outer mark and back. Marconi trailed the duelling *Shamrock I* and *Columbia II*, transmitting a steady stream of Morse code at 15 words per minute. The Navesink station received the messages and retransmitted them by telegraph to the *Herald*. A word tapped out on the *Ponce* arrived at the *Herald* seventy five seconds later. The *Herald* scooped all of its rivals.

So fervent was the interest in the wireless that reservations on the *Ponce* and the *Grand Duchesse*, both of which carried guests for the days of the race, were almost impossible to obtain. They were both luxurious ships that had been specially fitted out for this momentous occasion. Aboard the *Grande Duchesse* almost 1,500 passengers jostled for a view of the wheelhouse to witness the sparks as the messages were sent out. As the *Ponce* pulled alongside the *MacKay*, which was also in the parade, a woman guest aboard the cable ship picked up a megaphone and shouted towards the *Ponce*, 'Three cheers for Marconi!' The passengers aboard both ships took up the cry until the *Ponce's* Captain, also by megaphone, explained that Marconi was too busy sending messages to acknowledge the cheers.

At times the America's Cup International Yacht Race of 1899 struggled to produce any racing at all. It seemed that the reporting of the race by wireless was more exciting than the actual races as a serious lack of wind resulted in numerous 'drifting contests'. In total, eight race postponements took place due to lack of wind or to thick fog. On 8th October a start was given, but the time limit was reached and the race was cancelled. The America's Cup races were eventually held between the 16th and 20th October 1899, off New York, decided by the best three out of five races.

Unfortunately *Shamrock* was outclassed and during the second race on the first windward leg, *Shamrock's* topmast snapped and the challenger had to withdraw. The same evening a new spar was fitted and the Irish crew also added four tonnes of ballast. The yacht was tested on 18th October, but its sailing performance had not improved. It was found that overall the Irish yacht's steel mast was simply too flexible and during the 1899 yachting season the boat destroyed six sets of sails.

In the end the America's Cup stayed with the New York Yacht Club. Marconi's wireless system had covered every detail of the races. *Columbia* with Scottish-American skipper Charlie Barr and a handpicked Scandinavian crew had beaten *Shamrock I* by three wins to nil, but Sir Thomas Lipton's continual fair play brought unprecedented popular appeal to the sport and to his tea brand. Marconi and Lipton became great friends.

Following the competition, the *New York Times* published that the 'initial success of Marconi appeals powerfully to the imagination. It will be the fervent hope of all intelligent men that wireless telegraphy will very soon prove to be more than a 'scientific toy'.

Before the short demonstration to the U.S. Navy and the start of the race coverage, officials from the Navy Department had been inclined to be sceptical about the possible benefits of wireless communication.

After the races, comment in the Navy Department was optimistic and a Washington newspaper even reported Rear Admiral Bradford as being very much pleased with the reported success of the tests. US Navy Lieutenant Commander E.F. Qualtrough exuberantly commented to a *Herald* reporter:

'If we could only have had this last year, what a great thing it would have been. When we landed marines in Guantanamo, the ships were unable to lend assistance, for the reason that the enemy could not be located, and firing at random would have placed our forces in danger. With the aid of the Marconi system, the men ashore would have directed the fire.'

The subsequent trials with the U.S Navy had mixed results. The equipment performed well under difficult conditions. Prior to agreeing to the US Navy tests Marconi made it clear that the equipment provided for the races was short range and only sufficient for the distances required, and was not the type of equipment his Company normally fitted into naval vessels. He further stated that with the available equipment he would not be able to obtain results comparable with those obtained in the British Navy manoeuvres of the previous July. The U.S Navy were impressed by the results but concerned about the interference caused by two wireless transmitters in close proximity. Marconi was concerned about the risk of industrial espionage - during his America trip it seemed that everyone wanted to study his equipment and take it apart. So he had deliberately not taken his latest tuneable equipment to the United States.

It was time for Marconi to return to Britain, a country that was about to go to war. The Anglo Boer War was fought by Britain and her Empire against the Boers. The Boers were comprised of the combined forces of the South African Republic and the Republic of the Orange Free State. The Boer Republics declared war on 11th October 1899 and in November 1899 Marconi wireless equipment, manufactured at the new Hall Street works, was deployed to South Africa to serve with the British Army during the Second Boer War. It was the first time that wireless equipment would be used in a war.

The Marconi engineers, attached the Royal Engineers, Boer War

HMS *Thetis*, equipped with wireless to during the Boer War

The equipment was accompanied by a small team of Marconi engineers to install and operate the systems and train the Army personnel. The Marconi engineers were Mr. G.L. Bullocke, who had run the South Foreland station and was placed in charge of the Marconi team; he was accompanied by H.M. Dowsett, W.R. Elliott, C.S. Franklin, Carl H. Taylor and J.C. Lockyer. The shipment of five complete wireless sets and trained personnel travelled out on the SS (also RMS) *Servia* and arrived in Cape Town on 24th November 1899. The young Charles Samuel Franklin was on his first appointment for the Marconi Company having just graduated from University.

However the young men quickly found that the original agreement and assignment had been changed and they were now invited to *volunteer* for active service in the field. The men were prepared to do so, but the equipment, which had been designed and tested for fixed shipboard use, now had to be hurriedly installed in wagons for portable use over land. These may well have been the first fully mobile wireless system in the world, but the Marconi Company's vision of the best military application differed wildly from the military commanders'.

Wireless, the Company felt, should be installed behind enemy lines to report on enemy activity or be located at headquarters to direct overall troop movements. Instead, the now mobile wireless sets and the Marconi engineers were sent straight to the front line. The Marconi wireless equipment was originally intended to be deployed around *De Aar*, the railhead for the dispersal of the British forces. But the wireless telegraphy sets were now directed to provide communications between various British columns operating in the area although it quickly became clear that the wagons used for the mobile installations were unsuitable for the task.

The original unsprung wagons brought from England proved totally unsuitable for use in the South African bush, causing damage to delicate Marconi electrical equipment as well as the risks to the balloon envelopes during transportation. The Royal Engineers quickly built wagons to the Australian pattern, as used in the NSW Army Medical Corps for its ambulances, which featured sophisticated suspension. Three of these hastily assembled wireless stations, complete with Marconi engineers and R.E personnel, were then attached to Lord Methuen's column of 13,000 men marching to relieve Kimberley and Mafeking. The two remaining stations and personnel were sent to General Sir Redvers Buller's column of 18,000 men marching to relieve Ladysmith.

It was a baptism of fire for wireless telegraphy on the battlefield, but it did not get off to an auspicious start. The Marconi engineers were given very little time to set up and calibrate their equipment, nor any time to confer amongst themselves before being hastily sent into the field. They also had no time to train the Royal Engineers' personnel or personally adjust to the rigours of wartime service life. It quickly became clear that the harsh climate and conditions of war torn South Africa were to prove more than the embryonic communication system could cope with. The ironstone studded, dry sandy plains of the Northern Karroo in South Africa proved to be extremely difficult to cross with continual shocks, vibration, exposure to dust and extremes of temperature often rendering the equipment unusable. The bamboo poles soon began to split in the arid conditions. As the masts failed, kites and balloons were used in an attempt to provide the spark transmitters with suitable length aerial wires as the aerial length was crucial for tuning the system. When the wind was favourable and the kites flew well, communication was possible between the Orange River and the railhead at De Aar nearly 60 miles away although it was omitted from early reports that this required the use of a relay station at Belmont.

Almost constant static discharges from sand storms and cyclones continued to desensitise the equipment, while intense lightning storms across the South African veldt also had a paralysing effect on the coherers within the receivers. These storms were almost a daily event. They also wrecked the bamboo masts and the kites carrying aerial wires aloft that were difficult to synchronise at the two stations at the best of times. The 14 foot long balloons tried as alternatives were torn adrift and lost, the wagons continually broke wheels and axles and the Marconi engineers had to contend with inexperienced teams at both ends of any link. It is likely that no two wireless sets were ever operating on exactly the same frequency because of the variability in the height of the aerials.

Also the quality of the equipment's earth connection was seriously impaired by the nature of the ground. Captain Kennedy reported that the engineers tried to rectify the problem of poor ground connection by burying sheets of tin below the aerial masts, but apparently they achieved little success. Owing to the adverse weather conditions, the Marconi equipment remained unserviceable for three of the six weeks that were spent evaluating the system in the field. Unsurprisingly highly negative reports started to filter back to England which indicated that the Marconi wireless equipment was considered unserviceable and unusable in South Africa and the Director of Army Telegraphs ordered the sets to be dismantled and returned and ordered that two further sets, which had been sent to accompany General Buller's forces in Natal, were also to be withdrawn from service.

But the Army's curt dismissal of wireless was not echoed by the Admiralty, who immediately stepped in and asked to use the five Marconi wireless installations. At the turn of the century, the Admiralty and the Royal Navy would quickly evolve from wireless innovators to wireless consumers, but in both instances they displayed a strong sense of urgency and had always set the agenda. In the brief interim period around the turn of the century, the new conflict halfway around the globe offered up the first chance to test wireless communications in the field of battle.

The wireless apparatus was quickly adapted for use at sea and the Marconi engineers were instructed to install the equipment aboard the navy warships to support the blockade of Delagoa Bay. In this bay the town of Maputo (previously Lourenco Marques) is situated, and it offered the closest harbour facilities for importation of war material for the Boer forces. The five Marconi sets were installed on the cruisers HMS *Dwarf, Forte, Magicienne, Racoon* and *Thetis* by March 1900. With the permanent installation made possible by shipboard use, the equipment soon came into its own. The ships were able to cover much larger search areas, out of sight of each other, and still maintain communications, regardless of whether it was day or night, and in every conceivable weather conditions. Moreover, with HMS *Magicienne* lying at anchor in Delagoa Bay and serving as a relay station, it was possible to communicate with the Naval Commander in Chief in Simonstown by means of a combination of wireless telegraphy and landline. Simonstown is more than 1,800km from Maputo along the coast.

For effective use of the equipment, the masts of the ships were extended to accommodate the long wire antenna used. Given the ideal conditions for radio propagation by ground wave offered by the highly conductive sea water, it was possible to communicate regularly over long distances. On 13th April 1900 a range of some 85 km was reported to have been achieved. By November 1900 the war had entered a new phase, with the Boer forces resorting to Commando or guerrilla tactics for another 18 months. There was in effect no continued need for the naval blockade and by November the Marconi wireless telegraphy equipment was withdrawn from Royal Navy service and put into storage.

In part as a result of this wartime success, on Monday 2nd July 1900 a contract was entered into with the Admiralty for the installation of the Marconi apparatus on twenty eight Royal Navy ships and four coast stations. The Navy also agreed to keep and pay for the five Boer war sets. Marconi's bold step of investing in the Hall Street factory, plant, machinery and staff proved to be an act of inspired genius.

On 24th March 1900 the Wireless Telegraph and Signal Company was reconstituted to become Marconi's Wireless Telegraph Company Ltd. The name change was a Board decision, despite Marconi's objection. The Company considered that given Marconi's international fame, his very name was now a major asset. Even in the days before tabloid newspapers and paparazzi, the pressures of success were building on the young Italian's shoulders. Until this time, Marconi had been personally consulted upon all enquiries and decisions; he then gave his instructions and Managing Director Henry Jameson-Davis delegated members of the staff to deal with them. But the Marconi Company was now growing rapidly and it was becoming impossible for Marconi to be everywhere and for him to be involved in everything.

With the formation of the new Company Marconi's cousin, who had met him on the station a few short years past, Jameson Davis stood down (as he had always intended to do) and a new Managing Director, Samuel Flood-Page took the helm. He found himself in technical control of the growing Chelmsford Hall Street manufacturing works as well as having to deal with the increasing spate of technical inquiries. To share the load and delegate factory operations Flood-Page appointed a Hall Street works Manager, Mr H. Cuthbert Hall, who took up the post on 1st April 1901.

At the same time, Edmund Arthur Norman Pochin, a qualified industrial electrical engineer, assumed overall responsibility for the technical control of the Company with the title Chief Engineer. Pochin had trained at the Anchor Street Works of Crompton and Company in Chelmsford as an electrical engineer and later became manager of the firm's lamp department where he invented the first Crompton-Pochin arc lamp. Subsequently Pochin resigned from the Marconi Company and founded his own business and was awarded the M.B.E. in the 1919 New Year's Honours for services in the war manufacturing for rifle gauges. Andrew Gray, on returning from the United States, took over Pochin's duties with the title at Marconi's suggestion of Chief of Staff of the Marconi Company. The title reverted to Chief Engineer some years later.

In late January 1901, HMS *Jaseur* using Marconi's new tuned receiver successfully received signals from both the Haven Hotel wireless station in Poole and from HMS *Hector*. This is the first time at sea that two stations were received by a single vessel with one receiver. In 1901 a further 50 'service sets', known as 'Service Gear Mark II', but also referred to colloquially by their registration numbers as '1 to 52' sets, were ordered for fleet use. All these sets were manufactured at the Marconi Hall Street works in Chelmsford. By 31st December 1901 the Royal Navy had 105 wireless sets in operational use. Also in 1901 the Queensland Government bought two Service Mark II kits.

The Royal Navy were to lead the world in the adoption of the Marconi system, and Admiral John Fisher, First Sea Lord, was adamant that the new ships should have:

> 'No masts or fighting tops: only a pole for wireless. The necessity for masts and yards for signalling does not exist.'

The Royal Navy contract was further extended in 1903 and in 1904 the Royal Navy began to use the Marconi wireless system exclusively. By 1908, the importance of Admiralty wireless messages was acknowledged in the 'Handbook for Wireless Operators' which noted that distress calls had priority followed by Admiralty messages and then safety messages, also known as danger messages, which were preceded by the Morse code signal TTT.

Meanwhile the Royal Navy coastal wireless shore stations continued to be developed including 'Telegraph Tower ' on the Isles of Scilly as well as Culver Cliff, Dover, Portland, Spurn Head and Landguard [Fort near Felixstowe in Suffolk] in England, St. Anne's Head in Wales and Roche's Point and Bere Island in Ireland. The Admiralty also had hub stations in major locations such as Gibraltar & Malta with a central station in London called 'Whitehall Wireless.'

Within five short years the Royal Navy was, by 1905 totally dependent on Marconi wireless equipment, all built at Hall Street for its operations with over 80 ships fully equipped with long range telegraphy equipment. By use of the huge 'transatlantic' station at Poldhu Point, the wireless systems that grew out of the Boer War experience allowed the Admiralty to reliably communicate with any ship of the Royal Navy throughout the Atlantic Ocean, North Sea and Mediterranean.

In 1904, Brazil had begun a major naval building program that included three small battleships. Designing and ordering the ships took two years, but these plans were scrapped after the revolutionary new 'dreadnought' concept rendered the Brazilian design obsolete. The Brazilian Navy then ordered two new dreadnoughts to be built in the United Kingdom instead, making Brazil the third country to have ships of this type under construction even before traditional Naval powers such as Germany, France, or Russia. The Marconi Hall Street works supplied wireless equipment for both of the battleships, now called the *Minas Geraes* and the *São Paulo*. These two ships were intended to be Brazil's first step towards becoming an international power and soon initiated a South American naval arms race. While being built the ships created much uncertainty among the major countries in the world, many of whom incorrectly speculated the ships were actually destined for a rival nation. Similarly, they also caused much consternation in Argentina and Chile.

As the first Royal Navy contract started in 1900, naval dockyards often had several Marconi men fitting and operating ships' sets with others undertaking experimental work on a small scale at the coast stations. Marconi also had his loyal band of dedicated experimental engineers and constructors. These groups of young men all worked with energy and enthusiasm and with a small factory in full production at Hall Street in Chelmsford they formed the nucleus of a highly flexible organisation.

Marconi engineer M. Knight remembered that the apprentices regularly used to compete with each other by climbing up the Hall Street Factory's large aerial mast. New recruits to the staff learnt 'on the job' from more senior men but it was realised that a haphazard training of this kind was not very satisfactory. The men directly involved with experimental work were well taught, but those on shift duty or at isolated coast stations were not so fortunate. In those early days, however, there were more jobs than men to carry them out, so that this disparity in training was partly discounted by moving staff from one job to another every month or so. As the organisation continued to grow, men often remained for longer periods in remote stations without transfer.

The Royal Navy contracts were to provide a life line for Marconi's ongoing experiments to develop his wireless system. Hall Street was now in full time operation building equipment for the Royal Navy, Lloyd's of London's coastal stations, numerous merchant ships and soon a generation of great Ocean liners.

But by far the biggest demand on the time and resources and staff of the Hall Street works came when the factory was called upon to provide the wireless equipment and engineering support for Marconi's transatlantic experiment. To achieve the first ever wireless signal across over 2,100 miles Marconi had designed and built two huge stations at Poldhu Point in Cornwall and Cape Cod in Newfoundland. In addition he had built another station in Chelmsford to conduct tests with the Poldhu station in Cornwall. It had always been understood that Marconi's new wireless research station in Broomfield (off Pottery Lane, on land owned by the Christy family) on the outskirts of Chelmsford opened in August 1903 for tests with the Poldhu station's new 'T' aerial operating on a wavelength of 2,000 meters. However recent research indicates that the wireless station at Broomfield was operational by late 1901 in time for the transatlantic experiments. The Broomfield site was later to provide the Marconi Company with a research station to study wireless transmission with powers somewhere between the huge Poldhu station and the limited power of marine equipment. The station was headed by senior engineer Henry Joseph Round.

Early undated photograph c. 1903. Broomfield Wireless Station. H.J. Round possibly far left

H.J. Round   Round in later life, possibly at the Broomfield station

The Newsman newspaper reported on 09/11/1901:

WIRELESS MESSAGES

'Since the erection of the new pole near Messrs Marconi's Wireless Telegraphy Works at Chelmsford on Monday messages have been sent to and received from the Company's station at Frinton-on-Sea and also a station at Sheerness. To be expected that another pole will be erected at the signalling station on the Broomfield Road.'

On 10th January 1902 the *Essex County Chronicle* reported:

OVER-LAND
WIRELESS TELEGRAPHY
EXPERIMENTAL STATION FOR CHELMSFORD

'Experiments in long distance over-land wireless telegraphy are shortly to be commenced at Chelmsford by the Marconi Company. For this purpose four huge poles, each 110 feet high, are being erected in a field at the rear of Messrs. Bryan and Sons' Potteries in the Broomfield Road. The poles will be in the form of a square, and will support a large number of 'aerials' or attracting wires which receive the messages sent from other stations.

Practically no long distance work on land has been carried out by Marconi, so the result will be looked for with interest. It is expected that messages will be sent and received across London, to and from Poole, Dorsetshire. This will be a notable feat, as there are in London so many obstructions, such as electric wires, etc., which attract and weaken the 'air waves' which convey the messages.'

In Cornwall the huge Poldhu station occupied a fifty acre (200,000 m^2) plot with building work underway from October 1900 to January 1901 to a design by Marconi Company consultant John Ambrose Fleming. Marconi had convinced sceptical investors and his Board of Directors to put £50,000 into the project to try to transmit messages across the Atlantic Ocean. Marconi had already sent and received messages over 225 miles from the Crookhaven station in Ireland to a receiver station located 4.5 miles away from the main Poldhu station at the Lizard. Both stations were equipped completely with Hall Street manufactured equipment, proving that 'over the horizon' communications were possible.

On the bleak Cornish coast severe gales in September 1901 destroyed the enormous twenty mast circular aerial array built at Poldhu. All twenty pine masts snapped like matchwood and collapsed in a shambles of shattered timber and tangled wire. The aftermath of the storm left the station in total chaos, and the only good news was that miraculously there had been no loss of life. Disappointment swept through the Marconi Company ranks. The engineers said it would mean a postponement of three months or more to remove the wreckage and build anew. But under George Kemp's driving project management skills station testing was resumed on the 26th September using two 160 foot masts. 200 feet apart with a triangular stay, from them were suspended fifty four, 7/20 gauge (0.35 inch diameter (0.9 mm)) bare stranded copper wires, one metre apart to form an enormous fan. Marconi's great experiment was still possible.

A second huge wireless station of similar design to Poldhu had been set up at Cape Cod in America, 2,200 miles away. However, gales in November 1901 completely demolished this aerial as well and the receiving station now had to be moved to St John's, Newfoundland. Here Marconi could only lift his receiving aerial wire by using a kite.

In December 1901 at great risk to his Company, Marconi gambled everything on what he called 'the great adventure' or 'the great leap'. On 12th December 1901 he successfully received the Morse letter S, sent from the huge transmitter station at Poldhu Point in Cornwall while seated in a windswept tower in Newfoundland. The story will be forever famous. 'An uncertain kite flying in a giant storm and thought passed between' wrote one commentator.

At that moment the world changed.

Marconi had forced a wireless signal over 2,170 miles across the Atlantic Ocean, and for the next ten years Marconi and his company would struggle to develop the expensive and technically challenging transatlantic wireless service. Through it all the Hall Street works would be used to produce every piece of wireless equipment that could be manufactured in-house.

By now the Hall Street works was employing around 55 people, men and women working separately. The men worked on lathes, milling machines, drills and saws, driven by overhead shafts along with the carpenters building equipment frames and enclosures. On the upper first floor, women were employed in winding, insulating and lacquering the induction coils for the spark transmitters.

The Hall Street works ground floor manufacturing area was powered by a series of driven line shafts which powered each machine by a leather belt. This system was used extensively from the start of The Industrial Revolution until the early 20th century when individual electric motors were able to be fitted to each machine. The system of belts, pulleys and gears was known as 'millwork' with the line shaft kept in continual motion and each flat belt drive to any piece of equipment could be disengaged by manoeuvring the belt from a fixed pulley onto a loose pulley (the 'idler') using a simple lever - the origin of the phrase 'playing fast and loose'. Stepped pulleys provide three drive speeds for the machine tool depending on which pair of pulleys was connected by the belt.

The machine shop would have been a cacophony of working machines and slapping belts. The fixed pulley on the upper shaft was driven at constant speed by a belt from a common power source, either a water wheel, turbine, windmill or a steam engine. The belts were generally tanned leather or cotton duck impregnated with rubber. Leather belts were fastened in loops with rawhide or wire lacing, lap joints and glue, or one of several types of steel fasteners. Cotton duck belts usually used metal fasteners or were melted together with heat. The leather belts were run with the hair side against the pulleys for best traction. The belts needed periodic cleaning and conditioning to keep them in good condition. Belts were often twisted 180 degrees per leg and reversed on the receiving pulley to cause the second shaft to rotate in the opposite direction.

Pulleys were constructed of wood, iron, steel or a combination thereof. Varying sizes of pulleys were used in conjunction to change the speed of rotation. For example a 40 inch pulley at 100 rpm would turn a 20 inch pulley at 200 rpm. Shafts were usually horizontal, mounted overhead and made of rigid steel, made up of several parts bolted together at flanges. The shafts were suspended by hangers with bearings at certain intervals of length. The distance depended on the weight of the shaft and the number of pulleys. The shafts had to be kept aligned or the stress would overheat the bearings and could break the shaft. The bearings were usually friction type and had to be kept lubricated. 'Pulley lubricator' employees were required in order to ensure that the bearings did not freeze or malfunction.

These flat belt drive systems became popular in the UK from the 1870s, with the firms of J & E Wood and W & J Galloway & Sons prominent in their introduction. Both of these firms also manufactured stationary steam engines. Until the first decades of the 20th century coal fired steam engines had been the primary source of power and industrial energy. But they were also a major cause of what was known as 'smoke nuisance'. In order to provide power for small shops and light industry, specially constructed 'power buildings' were constructed. Power buildings used a central steam engine and distributed power through line shafts to all the leased rooms. Power buildings continued to be used even in the early days of electrification, still using line shafts but driven by an electric motor. It is not recorded if the Hall Street works was powered by steam as it had been as a silk mill, however by the late 1880s many industries, especially those close to town centres such as Hall Street had begun to popularise the use of gas, a cleaner and more efficient alternative for providing motive power than 'King Coal'.

On the 9th March 1912 the *Essex Newsman* newspaper, referring to that winters coal strike were able to report that the Marconi works (at Hall Street) had sufficient coal to continue manufacturing, but this have been used for heating.

A gas engine had been invented in 1859 by J.J.E. Lenoir, later developed by Alphonse Beau de Rochas and Nicolaus Otto, to occupy a niche in the market that the steam engine had not filled satisfactorily, providing power to small workshops. The great advantage of employing a gas engine was that it could 'commence as well as cease its action at a moment's notice', doing away with the 'wasteful' necessity of having to maintain a head of steam at all times to operate machinery that was often needed only intermittently. The 'silent Otto' was the most successful gas engine, with Crossley Brothers of Manchester building around 35,000 under licence between 1877 and 1900, exporting them all over the world. Two and three horsepower versions of the 'silent Otto' were demonstrated at the Smoke Abatement Exhibitions of 1881 and 1882.

With factory electrification in the early 1900s, many line shafts began converting to electric drive. In early factory electrification only large motors were available, so new factories installed a large motor to drive line shafting and millwork. After 1900 smaller industrial motors became available and most new installations, including New Street, used individual electric drives

The *Essex County Chronicle* reported on 17th August 1900:

> 'Ever since Mr. Marconi put up his pole for the purposes of wireless telegraphy at Chelmsford, the people of the town have walked about under the proud consciousness that they possessed the tallest pole in Europe. It seems, however, that, in a sense, Chelmsford has been outdone. The pole at the Essex county town is in three sections, the height altogether being 150 feet. The largest single pole, I read, has been erected at Ilfracombe. It is to be used for wireless telegraphy experiments between the Mumbles and Ilfracombe. The height is 116ft. 3in.; it is 17in. in diameter at the base, tapering to 3 1/2in. at the top. Its weight is nearly two tons. It has been placed at a depth of six feet in the solid rock.'

On 18th October 1901 the *Essex County Chronicle* reported:

> 'Essex people, among whom Mr. Marconi has established his works, will hear with special satisfaction that a new advance in wireless telegraphy is recorded.
>
> Mr. Marconi has succeeded in transmitting messages through the air for nearly 350 miles, which is almost double the maximum distance hitherto reported. Perhaps after all, Tesla's dream of flinging wireless messages across the Atlantic may be fulfilled even in our time. An arrangement has been completed with Lloyd's which will make all their stations receiving and transmitting agencies, and as the great steamship lines are adopting the new system the passengers on some of the longest routes – to India and Australia, for example – will hardly ever by out of touch of land.
>
> The whole Mediterranean Fleet either has been or very soon will be in actual permanent communication, every ship with every other, from one

end of the great inland sea to the other. It is reported that during the recent race between *Shamrock* and *Columbia* efforts were made by means of discordant electric waves to throw the Marconi signals out of order, but without effect.'

On 1st November 1901 the *Essex County Chronicle* reported:

'Among the plans passed or provisionally approved were a new test room at the Marconi Telegraph Company's Works. In regard to a proposed new mast, 150 feet high, to be erected on the land of the Marconi telegraph Company in Hall Street the Sanitary Committee were advised that the Council had no jurisdiction in the matter, either to agree to the erection of the mast or to disagree.'

The *Essex County Chronicle*, 1st November 1901 reported:

MARCONI'S POLE AT CHELMSFORD

'Operations had been commenced to erect a tall mast near Marconi's Wireless Telegraph Company's Works in Mildmay Road, Chelmsford, to replace the one blown down in a gale some months ago, but yesterday morning, when the pole was about to be raised into position, a telegram was received postponing the work. It is now the intention of the Company to erect the pole in a field close to Chelmsford if possible. [At the Broomfield, Pottery Lane station]. There had been considerable opposition to the erection of a second pole near the works by the occupants of surrounding houses, who feared another accident. Everything that engineering skill could do to ensure safety was done.

The new pole is 150 feet high, which is rather less than the previous one; it is in three pieces, the foot or lower mast being 76ft. long, the top mast 45ft. and the top-gallant mast 43 ft. Before, railway metals were driven into the ground for the wire and rope supports, but this time four anchor plates, 4ft. square, had been placed 7ft. deep, with three tons of concrete above each, with iron stirrups, for the steadying wires, which are of the best flexible steel. Altogether, over 12 tons of concrete were used. The accident to the first pole was owing to one of the galvanised straining screws opening out, but there will be no straining screws on the present mast, and each individual stay will bear a strain of at least five tons. The erection would have been as safe as it was possible for human hands to make it.'

Typical Marconi wireless station aerial mast - This at Marconi's South Foreland station- There is no known picture of the Hall Street wireless station

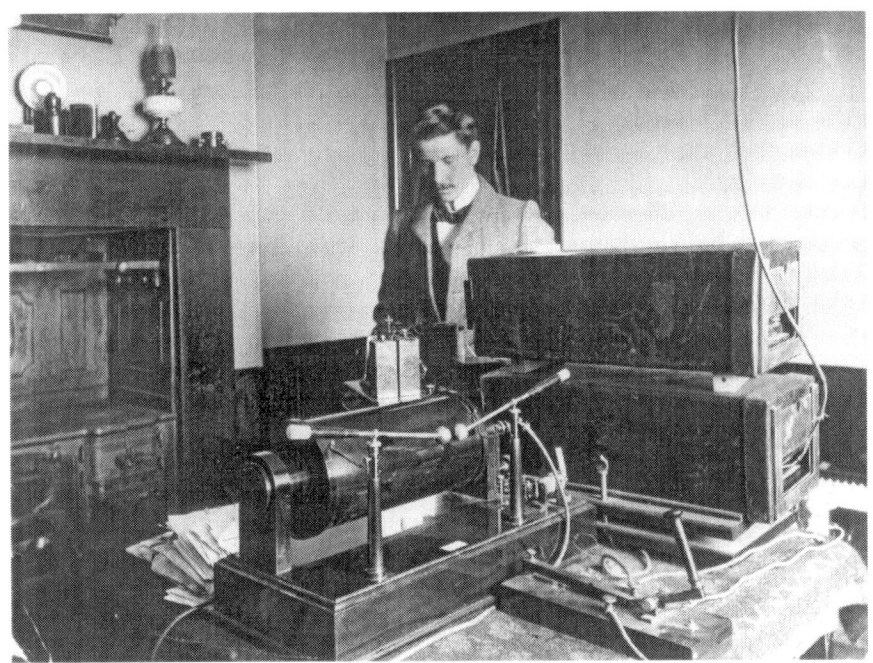

Typical Equipment manufactured at Hall Street for the South Foreland wireless station. Early Marconi engineer, G.L. Bullocke is pictured with - to the front the spark transmitter, with the induction coil creating a spark across the two spheres when the Morse key (bottom right) is depressed. The two long black metal boxes each house a sensitive coherer receiver with the aerial wire connected.

Marconi's Niton Station wireless room, c. 1899

Marconi wireless equipment onboard the *Tongue* lightship

On 15th September 1903 Captain H. B. Jackson, R.N., FRS., from HMS *Caesar*, Lieutenant (T.) C.R. Payne, R.N., from HMS *Vernon* and Lieutenant (T.) F.G. Loring, R.N., (Naval Reserves) visited the Marconi Company's Works at Hall Street in Chelmsford. They also visited the Marconi stations at Niton and Poldhu as part of the same inspection tour.

Their report to the Admiralty, dated 29th September gave some background to the equipment being built and the work underway at Hall Street:

> 'All Wireless Telegraphy instruments supplied by the Company are made at these works; the number of hands employed vary with the amount of work, but the average number is about a hundred, chiefly consisting of boys and women trained by the Company, considerable difficulty have been found in obtaining trained men. A tall mast, 180 feet high, stands in the grounds adjoining the works, the aerial wire from it being led directly into the Wireless Telegraph Office close to the foot of the mast.
>
> The earth consists of a number of galvanised iron plates (surface area of plates 1,000 square feet) buried vertically in the ground round the Wireless Telegraph Office. Great stress is laid on the importance of having the earth so that the leads from the receiving and transmitting instruments may be as short as possible, and of equal length. The Wireless Telegraph Station at the works is a small power one, and is used principally for experimental and testing purposes in connection with Marconi's station at North Foreland. The following are the names of the head officials at the works:- Mr. Grey, Chief of Electrical Staff, Mr. Priddle, Engineer in Chief. Mr. Ashley, Overseer and Instructor.'

They reported that the equipment being manufactured and assembled included:

> **Single Cell Accumulators**, made by the Chloride Company to the Marconi's Company design.
> **Special Induction Coil for Field Service Use.** This coil is enclosed in a strong wooden case, and thus protected from the weather. It appeared well suited for the purpose for which it was designed.
> **Magnetic Detector.** This instrument is the most sensitive receiving apparatus yet invented, but it is not suitable for ship work, as it depends on the receipt of message by telephone, and any noise in the vicinity prevents the signal being heard; also no permanent record is obtained, and great skill is required on the part of the operator.
> **Apparatus for testing Coherers.** The transmitting aerial, 3 feet in length, is placed at one end of a table and excited by a buzzer, the receiving aerial and coherer are placed at the other end of the table. A metal tube connected to earth can be lowered over the transmitting aerial, and so screen the waves from the receiver. By

the position of this tube the sensitiveness of the coherer can be gauged.

Insulators. A special form of porcelain insulator is used for the top of aerial wire. All wire stays for mast are insulated in three places by wooden dead eyes set up with hemp lanyards, which prevents signalling range falling off in wet weather.

Apparatus for reducing the Noise of Spark. The spark balls are enclosed in a glass tube with wooden ends, through the centre of which the rods carrying the balls pass. A small quantity of quicklime is placed in the glass tube to absorb any moisture which may collect on the inside of the tube.

New Pattern Receiver Box. This is a great improvement on the old pattern box. The size of the iron box and the wooden base board for instruments is the same as the old pattern, so that the old pattern may be easily brought up to date with the latest improvements.

The Newsman newspaper reported on 29/08/1903:

'The annual outing of the staff of Marconi's Wireless Telegraph Company took place on Saturday, to Harwich and Ipswich. Dinner and tea were served at Harwich. At the former repast 'The Health and Prosperity of the Firm' was drunk, as was also that of 'The Manager of the Chelmsford Station' (Mr. E. Priddle). The female employees went to London the same day for their outing. The arrangements were made by Mr. Piatanor and Mr. Priddle.'

As the Marconi Company rapidly developed its wireless communication system it became clear that the application of electromagnetic principles and engineering under many different circumstances which had never been encountered before, had created a whole new culture which had to be taught to the new recruits before they could be entrusted to any erection or installation work themselves.

On 25th April 1900, Marconi's 26th Birthday, The Marconi International Marine Communication Company Limited had been formed. Its aim was to develop reliable two way compact wireless systems for ship to ship, and ship to shore communication and it became evident that centralised instruction for new recruits in the technique of wireless telegraphy had become a necessity. A residential School for the training of probationer engineers was opened at Frinton-on-Sea on the Essex Coast in September 1901. This was the first wireless telegraph training college in the world, and it established a precedent in industrial technical training institutions.

The school occupied the building on the left of the photograph, while the other housed the school's resident students, with a 120 foot sectioned wood mast standing between the two. Just why Marconi came to Frinton is unclear, although a brief look at any map of this stretch of coast shows it to be readily accessible, with clear sea paths to North Foreland and the Kent coast. The majority of the students' time at the Frinton 'Wireless Telegraph Training College' was concerned with the use, maintenance and repair of transmitting and receiving equipment. It is thought that the station's local test transmissions were usually to a small (possibly portable) station located somewhere in Pole Barn Lane some 500 yards away. However the school's large mast would have enabled the students to easily communicate via wireless with the main Marconi factory in Hall Street, Chelmsford and with coastal stations at Dovercourt near Harwich and North Foreland over mainly sea paths. Hall Street produced all the equipment for the students and its experimental wireless station was able to operate with Frinton free of the requirement to handle commercial maritime traffic.

In later days a telegraph training college would usually train men to allow them to qualify for a Post Office licence as ship-to-shore wireless officers. However in 1901 marine wireless operators as a body did not exist and the Frinton School concentrated on teaching wireless skills to engineers selected from Universities and Technical Colleges. They were placed under the charge of a senior engineer for technical instruction, but were also allowed all the freedom necessary for independent study and experimental research.

The first Engineer in Charge was Thomas Bowden, whose experience with the Marconi Company dated back to 1898 when he had accompanied Marconi to the America's cup trials in New York. The work carried out at Frinton covered experimental studies of tuned and coupled circuits and work on high frequency transformers of the Tesla oscillation coil type, Morse code transmission and reception and Morse code inker tape operation, instruction in the erection of stepped 150 feet masts and experience in laying earth plates. Coherer testing was carried out between the school at Frinton and the North Foreland experimental station, coherer reception having reached its peak as earlier in the year Marconi had received signals on the SS *Philadelphia* at a distance of 2,000 miles from the famous Poldhu Point station in Cornwall. High Frequency testing equipment was almost entirely restricted to the use of hot wire ammeters and a point spark gap for the determination of transmitter tuning, and the strength of signals at a distance for testing of receiving equipment.

For text books the students used Maxwell's Electromagnetic Theory, Hertz's Electric Waves, J.J. Thompson's Electricity and Magnetism, supplemented by S.P. Thompson's and J.A. Fleming's books on general electrical theory, electric machines and transformers together with careful study of all Marconi patent specifications. Several of the students of the year 1902 went onto fill higher positions within the Marconi Company and its associated organisations at home and abroad, and the names of many of them are on the Veterans' Roll.

Frinton was closed in 1904 and the school was transferred to the Hall Street works. For a time all the students were absorbed directly in the factory personnel of the general research, development and testing areas.

It was a difficult commercial period for wireless. The magnetic detector had just replaced the coherer as the principal receiver component and small low power sets were now available for ships. However although the shipping companies recognised the value of wireless, they were generally not prepared to pay for its development, installation or ongoing maintenance.

Despite the reluctance of the ship owners to install wireless equipment Marconi proceeded on his quest to challenge the transatlantic cable operators. On the night of 18th January 1903, Guglielmo Marconi and his associates gathered at the Marconi wireless station near South Wellfleet, Massachusetts. A message of greeting in Morse code was sent from President Theodore Roosevelt to King Edward VII of England, received at the Poldhu station. The event made the front page of the *New York Times* as the first transatlantic wireless message from an American president to a European head of state and the first official transatlantic wireless transmission originating in the United States.

Transatlantic communication was now a proven fact. Just three months after this telegram, on 30th March 1903 *The Times* newspaper contracted with the Marconi Company for regular transmission of news from the New World to the old. On 22nd August 1903 the Poldhu station also began transmitting a regular news service for shipping in Morse code on a fixed wave length of 2,800 metres. In October, this service enabled the Cunard steam ship RMS *Lucania* to keep in touch with world news direct from shore throughout her entire crossing from New York to Liverpool.

In June 1904 a regular commercial service was established to transmit nightly news summaries to subscribing ships, who could then incorporate them into their on board newspapers. The first regular ocean newspaper, the *Cunard Daily Bulletin* was published on board the RMS *Campania,* printing news items received from Poldhu. Their introduction was reported in 'Mid-Sea Wireless Telegraph News', from the May 1904 issue of *The Electrical Age*. By late 1906 issues of the S.S. *Hamburg's* onboard newspaper, *The Atlantic Daily News*, featured news reports 'received by Special Marconigrams', and passengers were also notified that they could send telegrams to nearby ships and shore stations.

But there were still problems. All the congratulatory wireless messages and ships news services still did not mean that Marconi had reached his ultimate goal of a full commercial service, with paying customers, that would rival or exceed the transatlantic cable system. These first short messages often took hours to send and had to be repeated many times, because reception was extremely variable. The unpredictable variations in signal strength were caused by natural fluctuations in the ionosphere, but this was not understood at the time and sometimes message sending rates were frustratingly slow.

Experiments with the apparatus at the stations to overcome these difficulties extended over the next two years, during which time the press grew critical and investors became restless. These experiments included doubling the power input to the station from seventy five to one hundred and fifty kilowatts, increasing the size of the antenna by adding outlying poles, and increasing the wavelength used (i.e. decreasing the transmitter frequency). Although still insufficient to solve the overall problem, each of these step changes produced gradual improvements and indicated the directions in which Marconi should proceed. After two years of experimentation Marconi came to the conclusion that, in order to ensure reliable communication between North America and Europe, a much larger aerial system was required. The South Wellfleet, Cape Cod station had proved to be less practical for relaying transatlantic messages than the more northern Glace Bay station. However, it worked well as Marconi's main North American 'ship to shore' wireless station. The station was the Cunard liner *Lusitania's* direct link to America while out at sea. The passengers received daily news reports and sent personal wireless messages called *Marconigrams*.

In 1904 the Company decided to gamble on building two larger and more powerful stations to replace Poldhu and Table Head. The new sites were chosen at Clifden on the west coast of Ireland, and at a location just south of the Table Head/Glace Bay station that is now called *Marconi Towers*. During the winter of 1904/05, the Table Head Station was moved to a 32 acre site six miles inland and a much larger aerial system was erected, with the original aerial at the centre. The transmitter power was also greatly increased with much better results. Over the next two years successive increases in wavelength and power meant that reliable transatlantic communication was possible at night, and by the summer of 1905 reliable signals were recorded at Marconi Towers/Glace Bay from Poldhu with both stations in daylight. The Marconi Towers station occupied about five hundred acres of land required for the antenna arrays. In addition to the power house, transmitter building, and smaller operational buildings, there was a residence for the station manager and his family. Just after the station was built, Marconi and his new bride Beatrice moved into the residence together with the station manager, R.N. Vyvyan, and his wife. Beatrice was from Irish gentry and evidently found this arrangement quite confining in comparison to the manor house lifestyle she was normally accustomed to.

Marconi had found by trial and error that better results were obtained at longer wavelengths. He used a wavelength of 1,650 metres (about 182 kHz) at his first transatlantic station at Glace Bay, but still was confined to night-time operation. By June 1905 Marconi's equipment had been developed so that signals were now being received at Poldhu from the Glace Bay station with both stations in daylight on a wavelength of around 3,660 metres. While making improvements at Glace Bay, Marconi had recently devised a horizontal inverted L aerial which was found to have marked directional properties. However, as it was impossible to extend the aerial system on the headland at Poldhu, he decided to build a new station on the west coast of Ireland that would incorporate every new device known to the Marconi engineering team and take over the transatlantic service.

On 25th July 1905 Marconi and two companions travelled to Cashel, County Galway to inspect the Irish coast for a likely location. Shortly afterwards a site 20 miles from Cashel at Derrygimla, about 3 miles south west of Clifden was chosen and preliminary building work commenced in October of the same year. The Clifden site was 1.5 miles from the road to Ballyconneely at Ballinaboy and in close proximity to the large 10 acres 'Lough Enlaghnacourty', later often referred to as Marconi Lake. The first engineer in charge was Mr. Entwistle who was succeeded by Mr. Mathias who remained in charge until the station closed down.

The new Marconi Towers station was built in 1905 and the Clifden station was completed in 1907. At Clifden the new aerial system was the new directional type, aligned with Canada and designed to operate between 42-45 kHz (a wavelength quoted at 6,666 metres). It consisted of eight wooden masts each about 210 feet high on which the aerial wires were suspended. The masts originally consisted of three wooden poles strapped and bolted together; however improvements were made from time to time with a major alteration and the addition of four steel masts in 1918. The unsatisfactory glass plate condensers were replaced with huge air condensers at both Clifden and Glace Bay which required the construction of large buildings on both sites. The giant condenser was housed in a galvanised iron clad building 350 feet long by 75 feet wide with a height of 33 feet at the eaves. It used air instead of the previously used glass between the plates. To obtain the required capacity of 1.8 microfarad, 1,800 sheets of galvanised iron each 30 feet by 12 feet were required. These sheets were suspended 12 inches apart attached at the top to porcelain rod insulators. When completed the condenser was tested up to 150,000 volts. The earth system consisted of two sheets of heavy copper gauze, 600 feet long and 4 feet wide, buried in the ground in line with the aerial. About 200 feet of these strips were also laid at the bottom of the lake. There were also a number of steel wires buried in the ground and connected to the steel frame of the condenser house.

The steam turbines to drive the generator and the generating plant itself were completely new, with an output of 300 kW dc at 20 kV. This charged a bank of 6,000 cells (each 40 amp/hour) giving an operating potential of 12 kV, or 15 kV if the batteries were used in conjunction with the generator. The stacks of the six steam engines are clearly visible in photographs taken in 1912. These were provided with steam by peat fired boilers fed by a specially constructed Marconi light railway that brought in peat from a bog 1.5 miles away and coal and supplies from the road at Ballinaboy. The Marconi railway had three locomotives and two rail cars on a track gauge of two feet.

The Clifden wireless station   The Clifden wireless station condenser house

During the summer and autumn of 1906 the huge Irish station grew and when completed was described as a shanty town, the whole 300 acre site surrounded by a high barbed wire fence apart from where natural obstacles acted as boundaries. The entrance from the railway had a wooden sentry box with guards day and night. At its busiest period the Marconi station employed a permanent staff of 150 (including 10 engineers and 25 operators) with a further casual workforce of around 140 people with 70 men working full time to dig the peat.

The station operated 24 hours per day and the staff, all male apart from a few housekeepers, worked three 8 hour shifts. One of the most famous members of staff was Jack Phillips, who later died as the heroic chief Wireless Operator on the *Titanic*. The receiver house was located across the lake from the generator house in a disused quarry. The operators' bungalow, the engineers' houses and the canteen were located nearby. The receiving aerial originally consisted of two wires, each about 2,100 feet long, supported at the tops of the masts of the transmitter aerial. Subsequently these were replaced by four wires, each 4,000 feet long and the average strength of the received signals was greatly increased. The lead-in wires from the receiving aerial were carried on wooden masts to the nearby receiving house.

During the next two years Marconi and his wife travelled back and forth between Great Britain and Canada while he endeavoured to make the transatlantic service operational travelling from site to site and station to station just has he had done since the first trials on Salisbury Plain in 1896. During their last stay at Marconi Towers the station conducted tests by day and night beginning on 15th October 1907, and some 10,000 words were faultlessly exchanged, to the great relief of the Marconi engineers.

The inaugural message on the 15th was sent at 11.30 a.m. from Lord Avebury to *The New York Times*. After further experimentation and improvements a regular transatlantic wireless-telegraph service was finally begun on at 9 a.m. on Thursday 17th October 1907 between Clifden in Ireland and Glace Bay. This connected Europe with North America without cables, and opened for public use with the transmission of 10,000 words. The station used a wavelength of about 5,000 metres (60 kHz) which provided reliable daytime communications and usable, but more variable, night-time communications.

The service between Clifden and the Marconi Towers, Glace Bay stations was the first point-to-point fixed wireless service in the world and represented the culmination of over ten years of intensive effort and experimentation. The Hall Street works had been an integral part of both stations manufacture. The transatlantic wireless service was an early success; 100,000 words were carried within three months of its initial commercial launch. Customers were not entirely satisfied, however, since messages were subject to delay, with the Company blaming congested landlines serving the coast stations at either end. R.N. Vyvyan, the engineer in charge of the Marconi Towers station, wrote:

'Only those who worked with Marconi these (past) four years realise the wonderful courage he showed under frequent disappointments, the extraordinary fertility of his mind in inventing new methods to displace others found faulty, and his willingness to work, often for sixteen hours at a time when any interesting development was being tested. At the same time the Directors of the Marconi Company showed wonderful confidence in Marconi, and courage in continuing to vote the large sums necessary from year to year until final success was achieved.'

In 1907 both Glace Bay and Clifden were also equipped with a new form of disc discharger which overcame many of the problems of the old spark transmitter at Poldhu. Initial tests by both day and night began on 15th October 1907 and some 10,000 words were faultlessly exchanged to the great relief of the Marconi engineers. The inaugural message on the 15th October was sent at 11.30 a.m. from Lord Avebury to *The New York Times*.

At 9 a.m. on Thursday 17th October the first official messages came in from Glace Bay. These 'Marconigrams' or 'Aerograms' were charged at 5d per word for ordinary messages and 2.5d for press reports; this was half the charge of the transatlantic cable system. Reception rate was reported to be 30 words a minute in Morse code. The term 'Marconigram' continued to be used worldwide long after Marconi's had relinquished any control over the service.

The unlimited service began in February 1908 and in the first five months over 68,000 words in the form of 'Marconigrams' were efficiently transmitted for *The New York Times*. So useful had wireless become to the paper that it boasted: 'The first and only newspaper to use the transatlantic wireless telegraph, by which it receives daily more than 2,000 words from Europe.'

The despatches were all marked, 'By **Marconi Wireless Telegraph.**'

The transatlantic service was an early success; 100,000 words were carried within three months of its initial commercial launch. Customers were not entirely satisfied, however, since messages were subject to delay with the Marconi Company blaming congested landlines serving the coast stations at either end. The Company responded by announcing plans for upgrading Poldhu and Cape Cod to relieve pressure on Clifden and Glace Bay and to run in parallel with these stations. The four stations, it estimated, would then be able to process 20 words a minute, 12 hours a day, generating a net annual revenue of £150,000.

Marconi's dream of building wireless communication across the Atlantic had come of age, and much of it had been built at the increasingly overloaded Hall Street works, but at last it seemed as if he might be able to challenge the cable operators.

But it had been a cripplingly expensive process for his young Company.

The Marconi Hall Street works in Chelmsford

Marconi and the Hall Street Engineers
Rear L-R: T. Bowden, J. Dolton, J. St Vincent Pletts, D.E. Newman, R.N. Vyvyan, E.T. Priddle,W.E Eccles, C.E. Rickard. Front Row: G.S. Kemp, G. Marconi, J. Erskine-Murray

Hall Street, The Ladies of the Winding Shop

Hall Street, The Ladies of the Winding Shop

Marconi's Hall Street works, Ground Floor Machine Shop

Marconi's Hall Street works, Ground Floor Machine Shop

Marconi's Hall Street works, Ground Floor Machine Shop

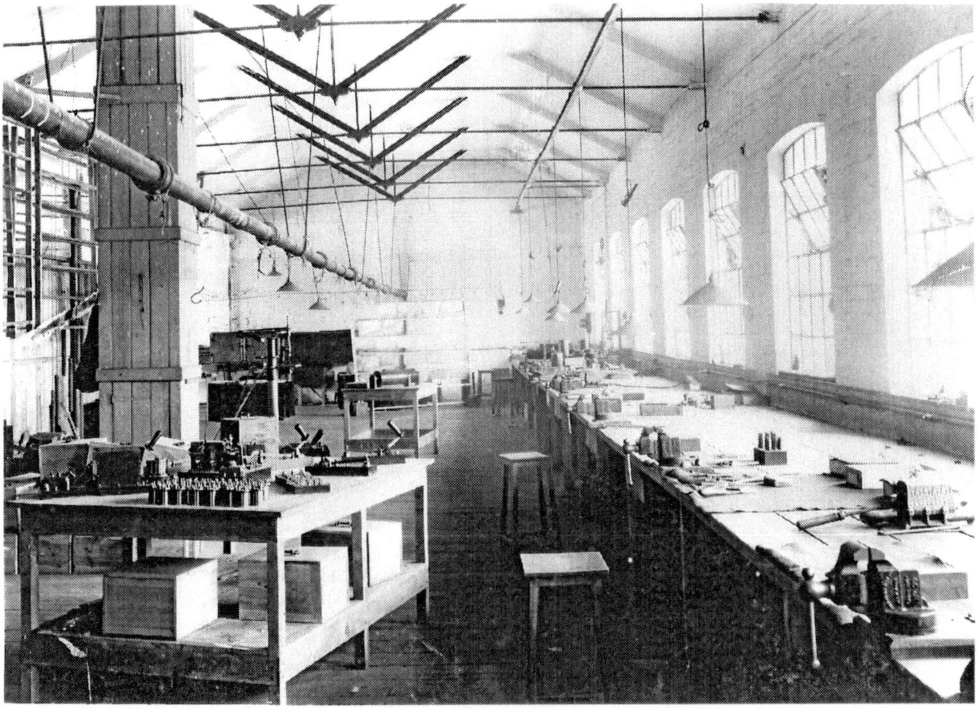

Hall Street, Top Floor Research Laboratory

Hall Street, Coil Drying Shop

Hall Street, Coil Mounting Shop

CHAPTER THREE

A Brief Sojourn to Dalston and the *Republic* Accident

Dalston Works, c. 1903

Dalston Works, c. 1905

In 1905, Cuthbert Hall, who had taken over as the Marconi Company's new Managing Director decided that the Hall Street works was overloaded and a new larger factory was required that should also be located closer to London.

The Dalston Street Works in Tyssen Street, North London, was opened by Marconi's in 1905 and occupied the Shannon furniture factory, an impressive building (now called Springfield House) that had been built in 1903-5 by Edwin Sachs. Its steelwork was encased in concrete and was brick-faced with heavy eaves. It was much larger than Hall Street, the property then had a total floor space of 31,170 square feet, valued at £61,585 in 1908, and included 'electrical power installation, lighting, heating and ventilation plant as well as fire appliances.'

The Marconi Dalston Works. Carpenters Workshop

The four storied Dalston Works building had three wings and was equipped with all the machinery removed from the Hall Street works in Chelmsford. It took over the role of Hall Street in the manufacture of coils for the Marconi spark transmitters and paper capacitors. It later began mass production of ignition coils for use in the expanding automobile industry.

The Marconi Dalston Works. Ignition coil department where girls work on the manufacture of mounting boxes. The machine shop has a single line-shaft supplying the drills, while the presses are hand operated

The Marconi Dalston Works. Capacitor Construction

The Dalston Works, 2013

As the Company's main production transferred to Dalston, the Hall Street works was not closed down as the Company had a long lease and was looking for additional space for its growing research divisions with brilliant engineers such as Henry Joseph Round and Charles Samuel Franklin playing the lead roles. Some of the earliest research into telephony transmission (speech rather than Morse Code) was to be undertaken at Hall Street.

Round had joined the Marconi Company in 1902 and was sent to the United States where he worked at Babylon, Long Island. In his spare time he constructed one of the first 'arc' *radio* telephones, in 1906. From a small transmitter near the Battery, New York, speech and gramophone records were transmitted to various places in New York including the Times Building and ships in the docks. On returning to England, Round continued with his 'arc' experiments and several sets were made in an attempt to provide a means of commercial communication.

These sets were used both in England and Italy and incorporated an arc that burned in an atmosphere of hydrocarbon vapour enclosed in a cylindrical chamber fitted with a mica observation window. One of the electrodes was kept turning by a clockwork mechanism in order to improve the steadiness of the burning. A gas reservoir was provided. The transmitter was fitted into a mahogany case and finished with ebonite panels and brass fittings. The 1910 transmitter/receiver never went into production, being overtaken by the ongoing and rapid research into the use of thermionic valves. But it did have a brief moment of glory. A test was arranged between two of the sets, one at the Hall Street works and the other at the Broomfield station a few miles away. Marconi himself travelled to Broomfield to speak over the wireless system and heard

Marconi engineer C.E. Prince (later Major Prince) replying from Chelmsford.

In March 1908 Marconi himself had to step in to replace Cuthbert Hall as Company Managing Director and under his direction the original factory at Hall Street in Chelmsford, which had been closed three years earlier, was re-opened and all the machinery and staff were transferred back to Hall Street. The large Dalston Works had been losing money and was closed down in 1908.

The year 1908 was to prove very difficult for both Marconi and his Company. The 'International Convention on Wireless Communication at Sea' had just come into effect which essentially ended Marconi Marine's near-monopoly on supplying wireless equipment for merchant vessels.

Ship owners could now buy equipment purely on its technical merit, from any supplier, yet too many of them still failed to do so. Further, the transatlantic wireless message service using the huge new Marconi stations that had taken a fortune to develop was still not yet profitable.

Marconi's had been in existence for twelve years and despite Guglielmo Marconi's many successes and world-wide fame, his Company was at a very low ebb. Over £500,000, nearly the entire cash reserve had been sunk in experimental work and no share dividend had yet been paid on the ordinary or preference shares and none was in sight. During 1908 the price of the Company's ordinary shares had dropped to 6s 3d and holders and investors who had paid £3 or £4 per share were not in a mood to make any further cash injections.

With his usual skill Marconi started to turn the fortunes of his Company around, and by the end of June 1909 the Hall Street works were busy once more with orders in hand worth £87,000 (today worth around £8 million). But Marconi had always found routine executive responsibilities extremely irksome. Nonetheless he had no alternative but to pursue an active policy of retrenchment after taking over as Managing Director to save the Company from the financial crisis that threatened to engulf it.

Part of the growth in orders undoubtedly came about due to the first use of wireless to save lives at sea when the RMS *Republic II*, a steam-powered ocean liner built in 1903 by Harland and Wolff in Belfast was lost at sea in a collision in 1909 while sailing for the White Star Line. Known as the 'Millionaires' Ship' on account of the number of well-known and immensely rich Americans who travelled by her, she was one of the largest and most luxurious liners afloat and was frequently referred to as a 'palatial' liner.

In the early morning of 23rd January 1909, while sailing from New York City to Gibraltar and Mediterranean ports with 742 passengers and crew and Captain Inman Sealby in command, the *Republic* entered a thick fog off the island of Nantucket, Massachusetts. Taking standard precautions and maintaining her speed, the steamer regularly signalled her presence in the outbound shipping traffic lane by whistle. At 5:47 a.m., another whistle was heard and the *Republic's* engines were ordered to full reverse, and the helm put 'hard-a-port'. Out of the fog, the Lloyd Italiano liner the SS

Florida appeared and hit the *Republic* amidships on her portside, at about 90 degrees. Two passengers asleep in their cabins on the *Republic* were killed along with another passenger when *Florida's* bow sliced into her, On *Florida*, three crewmen were also killed when the bow was crushed back to a collision bulkhead. Six people died in total.

The engine and boiler rooms on *Republic* began to flood, and the ship listed. Captain Sealby led the crew in calmly organizing the passengers on deck for evacuation. Republic was equipped with the new Marconi wireless telegraph system built at the Hall Street works and became the first ship in history to issue a CQD distress signal, sent by Marconi wireless operator, Jack R. Binns, whose messages were to save over 1,500 lives.

After the collision the *Florida* came about to rescue *Republic's* complement, and the U.S. Revenue Cutter Service cutter *Gresham* responded to the distress signal as well. Passengers were distributed between the two ships, with *Florida* taking the bulk of them, but with 900 Italian immigrants already on board, this left the ship dangerously overloaded. The White Star liner *Baltic*, commanded by Captain J. B. Ranson, also responded to the CQD wireless call, but due to the persistent fog, it was not until the evening that *Baltic* was able to locate the drifting *Republic*. Once on-scene, the rescued passengers were transferred from *Gresham* and *Florida* to *Baltic*. Because of the damage to *Florida*, that ship's immigrant passengers were also transferred to *Baltic*, but a riot nearly broke out when they had to wait until first-class *Republic* passengers were transferred. Once everyone was on board, the *Baltic* sailed for New York.

At the time of *Republic's* sinking, ocean liners were not required to have a full capacity of lifeboats for their passengers, officers and crew. It was believed that on the busy North Atlantic route assistance from at least one ship would be ever-present, and lifeboats would only be needed to ferry all aboard to their rescue vessels and back until everyone was safely evacuated. This scenario fortunately played out flawlessly during the ship's sinking, and the six people who died were lost in the collision, not the sinking.

RMS *Republic II*   Marconi wireless operatorJack Binns

Marconi went on to personally offer Jack Binns the position of wireless operator on the White Star's newest liner, RMS *Titanic* which was just being fully equipped with Hall Street manufactured wireless equipment. Jack Binns received a special medal for his services and Marconi himself presented him with a gold watch. But by this time the young 'Marconi man' was engaged, and his American fiancée didn't want him to return to sea. And so, he refused the job and one day before the *Titanic* sank, Binns began work as a reporter for a New York newspaper.

Captain Sealby and a skeleton crew remained on board *Republic* to make an effort to save her. Crewmen from the *Gresham* tried using collision mats to stem the flooding, but to no avail and on 24th January, the *Republic* sank. At 15,378 tons, she was the largest ship to have sunk up to that time. All the remaining crew were evacuated before the ship was lost.

The Newsman newspaper, 2nd October 1909 reported on a recent visit to Hall Street:

A VISIT TO MARCONI'S 'WONDER AND MYSTERY' WIRELESS AND THE 'WARATAH'

'On Wednesday afternoon members of the United Wards Club of the City of London, a society whose object appears to be to promote sociability among its members and meetings for intellectual recreation, visited Marconi's wireless telegraphy works, and were initiated into that wonderland of mystery; and in the majority of cases the visitors, as remarked by the Chairman of the United Wards, came away as wise as they were before the visit, albeit imbued with a sense of the wonderfulness and mysteriousness of it all. The party arrived by train at Chelmsford from Liverpool Street at about three o'clock, and were conveyed in brakes to the works of the Marconi's Wireless Telegraph Company, Ltd., situated in Mildmay Road. Here they were received by Mr. Dowsett, chief of the staff, Mr. Mitchell, works manager, with their assistants, Messrs. R. Cave, R.T. Munson, Ball, and Herring. The visitors were taken in parties to the various test-rooms and workshops, the latter presenting a hive of industry which was a surprise to many.

Recently the works have been extended, a gratifying sign both for Chelmsford and the Company, and in these new works some beautiful, not to say wondrous, wireless telegraphy instruments were shown and lucidly explained by the gentlemen named above. Here was seen working a 200-miles ship set, similar to the one on the '*Republic*' operated by Jack Binns, surely one of the heroes of the Marconi system. There was a working 400 miles set with wide range of wavelength, and a musical spark set capable of working even greater distance.

A similar set is now being fitted to the new Brazilian Dreadnought, the '*Minaes Geraes*', recently launched on the Tyne. The motors, dynamos, transmitters, converters and other equipment are simply a maze of the scientific construction, which mystifies the uninitiated. Lying dormant, the battleship set was impressive; in operation it was startling. The instruments seem to have harnessed the lightning, and, in launching forth their waves of fiery fluid, emitted a sharp, angry rattle, which seemed the very incarnation of force. Long and short flashes, which constitute the Morse code, rattled and flashed out their message en route to their ethereal pathway. And again was exhibited a 300 miles merchant service set, so compact that all the transmitting apparatus can be fitted into a silence cabin four feet square.

But the *mullum in parvo* [much in a small space] of the Marconi system is undoubtedly the portable Army set. The energy for working the dynamo of that set is supplied by one or two men working a pedal gear. It can be erected or dismantled in five minutes, and, including the mast sections, packs comfortably on the backs of two horses. Both sets have been supplied to Continental armies.

In other test rooms, demonstrations of accurately measuring electrical quantities with the most up-to-date instruments were given, also the calibration and testing of the smaller apparatus used wireless telegraphy. Then were shown magnetic detectors, which cause the waves generated by a distant station to produce suitable sounds; tuners, which enable neighbouring stations to work without interference; wave-meters and direction finders, etc. and last, but not least, an X-ray apparatus. Truly a wonderful exhibition, and all pulsating, so to speak , with electricity.

So far is recounted the finished works of the Company, but a word must be said of the works where the wonderful equipment for controlling one of the greatest forces of the age is made. The ground floor of the factory is a labyrinth of costly machinery, including lathes, drilling and other machines, some turning out the most delicate parts of wireless equipment. On an upper door are scores of girls making sections, etc. the whole affording an example of industry and methodical arrangement, upon which the responsible officers of the Company are to be congratulated.
But the great masterpiece of the exhibition and demonstration took place at the Broomfield Road station, where Marconi's system is in active operation. Here the great possibilities of the science were shown in other ways than for messages. For a flag was hoisted, and three detonators were exploded by wireless telegraphy, and one is made to shudder at the suggestion that it might be possible for, say, an enemy's ship off the Essex coast to send an electric wave to Waltham Abbey, and blow the powder magazines there to atoms. Another interesting demonstration was the exhibition of a working ship set in communication with (the Marconi Coastal station) at Caister, near Yarmouth, eighty miles distant, to and from which signals can be sent and received. The dispatching of the message was watched with absorbing interest. The electric fluid, when set in motion by the tap, tap, tap of the Morse instrument, cracked, rattled, and flashed with terrible energy near the dispatching instrument and then again at the summit of the tall mast, conspicuous for some distance around, the noise of the message as it leapt from the mast and launched into space could be distinctly heard. The message sent was from

the President of the London United Wards Club to the Lord Mayor of London, the message travelling via Caister. It was As follows:- 'Having a very pleasant afternoon, at Marconi's. President, United Wards Club.' The operator was Mr. R. K. Rice.

The apparatus at the Broomfield station also includes emergency gear, such as would be used in case of break-down in the ship's engine room; a working cavalry set, in communication with the Army set at the Company's Works; 300 mile Merchant Set, giving another example of a musical spark ; and a demonstration of wireless telephony, signals being sent from the main building and received on several portable sets while they were being carried about the field. Tea, supplied by Messrs. Hicks, Son, & Co., was then served in a huge marquee, and, at the conclusion, the President of the visitors, Mr. R. Riches, moved a cordial vote of thanks to Mr. Marconi for his kind invitation to view the Works. They had all thoroughly enjoyed the exhibition and demonstration, and were impressed with the wonderful strides made in wireless telegraphy since its inception, and, although many of them were little the wiser for the visit, it appeared to all of them both wonderful and mysterious. [*Applause*]

He concluded by paying a high tribute to the Marconi staff, who had been indefatigable in their efforts to make the visit a success. The proposition having been seconded and carried with acclaim, Captain Sankey, a director of the Company, responded. He apologised for the absence of Mr. Marconi and referred to the commercial importance of wireless telegraphy, speaking of the wreck of the *Republic,* when, by the warning sent out by the wireless telegraphy on board, every soul was saved. [Applause.] He expressed the opinion that if the *Waratah* had had wireless telegraphy aboard, that vessel would be been discovered.

[The SS *Waratah* was a 500-foot (150 m) long **steamship** that operated between Europe and Australia in the early 1900s. In July 1909, the ship, en route from **Durban** to **Cape Town**, disappeared without trace with 211 passengers and crew aboard.]

The party, numbering 150, then left for London. Others of the Marconi staff giving information to the party were Mr. E.H. Montagu, who was in charge of the Cavalry set Messrs. M.B.V. Dewar and R.H. White, who were operating a 5 K.W. Italian battleship set, Italian land and Mercantile Marine sets, and a 1.1/2 K.W. Torpedo Boat Destroyer set.'

Another headline news story in 1910 about wireless and the Hall Street factory surrounded the case of Dr. Hawley Harvey Crippen, better known as Dr. Crippen. He was an American homeopath, ear and eye specialist and medicine dispenser and became the first suspect to be captured with the aid of wireless telegraphy as he fled across the Atlantic, the equipment all manufactured by the Hall Street works.

The Crippen story aroused enormous public interest at the time. He was a small man, 5ft 4in tall, bespectacled and moustachioed, quiet, mild and polite but he had a wife who was described as a heavy-drinking nightmare, vain, bullying and promiscuous. The Crippens moved to London in 1897, where, while working at a centre for treating the deaf he struck up a relationship with one of the typists, an attractive girl called Ethel Le Neve. Ethel became his 'wifie' and he was her 'hub'. On 1st February 1910 Cora Crippen vanished. Her husband said she had gone back to the United States for a few months. In March Ethel moved into the Hilldrop Crescent house with Crippen who now spread the story that Cora had actually died in America.

Cora's friends grew suspicious, Scotland Yard was alerted and Detective Inspector Walter Dew interviewed Crippen in July. Crippen took fright and fled to Brussels with Ethel, who was dressed as a boy. The police searched the house and discovered the gruesome remains of a body beneath the coal cellar. For the press it became a cause célèbre. The Police watched ports and stations and police forces abroad were alerted. On 20th July Crippen and Ethel sailed from Antwerp for Canada aboard the SS *Montrose*, a transatlantic ocean liner of the Canadian Pacific Steamship Company. He called himself Robinson and Ethel again posed as his teenage son, but the captain grew suspicious of their behaviour and informed the ship's owners by wireless telegraph. They contacted Scotland Yard and Inspector Dew immediately pursued the fugitives across the Atlantic in a faster liner, the SS *Laurentic*. As the *Montrose* docked, Dew was waiting, went aboard with the pilot vessel and arrested Crippen and Ethel. He afterwards said that he had never in his life felt such a sense of triumph and achievement. The *Montrose* took all of them on to Quebec where hordes of reporters swarmed on board.

Back in London, the trial at the Old Bailey in October again stole the national headline and also made the name of the pathologist Bernard Spilsbury. He had examined a piece of flesh and confirmed that a scar on it corresponded to an operation that Cora Crippen was known to have had for the removal of her ovaries. Ethel was tried as Crippen's accomplice but was acquitted. Throughout the proceedings and at his sentencing, Crippen showed no remorse for his wife and concern only for his lover's reputation. After just 27 minutes of deliberations, the jury found Crippen guilty of murder. He was hanged by John Ellis, assisted by William Willis, at 9 a.m. on 23rd November 1910 at Pentonville Prison, London. Wireless, and Hall Street had joined the age of criminal justice.

As Hall Street gained new orders for marine equipment by 1910 it was now clear that the Hall Street works was simply too small as the demand for his new wireless telegraphy equipment had already increased tenfold. The Hall Street works were becoming seriously cramped. Double shifts were being worked in the factory to make wireless equipment for export to the four corners of the globe. Customers included the Amazon basin, Brazil, Thailand, South Africa, India and even to both sides in the Balkan War of 1912. In October 1911, the training School which had come to Hall Street when Frinton-on-Sea was closed down was re-established as a separate department at the Broomfield Research Station in North Chelmsford under R.G. Kindersley. (On 14th May 1943 two parachute mines fell flattening the Marconi Pottery Lane factory in Broomfield which consisted then of nine variously sized single storey buildings sited in the centre of a 10 acre field accessed from Pottery Lane.)

Even the transatlantic cable companies, Marconi's greatest competitors, became customers. Marconi wireless equipment was also the cornerstone of the growing number of shore based wireless stations and his equipment, all built at Hall Street, was carried aboard all the great Atlantic liners including the *Lusitania, Mauretania, Baltic, Olympic* and the ill-fated *Titanic.*

On 16th December 1910 the Essex County Chronicle reported:

A HUGE POLE

'A steel mast, 180ft. high, with a wooden top mast carrying to a height of 210ft., has been erected close to the Marconi Wireless Telegraphy Works in Mildmay Road. Some twenty tons of cement were used for the beds of the mast and supports. The pole is the tallest now used for wireless telegraphy, although one of 250ft. was formerly in use in South America. The present structure will be used mainly for experimental work with high power, and it will be easy to communicate with the transatlantic station at Poldhu, in Cornwall.'

By the end of 1910, Marconi was able to prove beyond doubt that wireless could provide an alternative to long-distance cables between fixed points. With 600 patents on equipment installed in a total of more than 500 wireless stations, its reach was now almost worldwide.

But Marconi was also tiring of the financial side of the business and continually sought to find a replacement Managing Director that would free him from the day to day operational decisions of the large Company, and allow him to return to his first love of pure research and development.

Godfrey Isaacs then appeared on the scene. Just turned forty, he was one of the younger sons of the large family of Joseph Isaacs, a fruit-broker and general merchant with a large and profitable business. Godfrey had no obvious qualifications for the post Marconi needed to fill since he had no knowledge of wireless telegraphy or the new industry. But his father's business, which he had entered as a young man, maintained links with exporters on the continent, and Godfrey, who had been educated at Hanover and Brussels University, spoke several languages and had extensive contacts within many of Europe's finance houses. He had tremendous energy and enterprise, and Marconi, to whom he was introduced by the inventor's brother-in-law, seems to have taken an immediate liking to him.

Godfrey C. Isaacs became joint Managing Director with Guglielmo Marconi in January 1910. In August, having proved himself to Marconi and the Company Directors, Isaacs took over as sole Managing Director. Isaacs was, like Marconi, multi-lingual and international in his outlook and he immediately undertook a full review of the Company's structure and operations.

In October 1910, Isaacs formed 'The Marconi Press Agency Ltd', a subsidiary, which in 1911 produced the world's first wireless magazine, *The Marconigraph*, renamed *The Wireless World* in April 1913.

The *Chelmsford Chronicle* on 14th April 1911 (announcing first copy of the in-house Marconigraph journal) reported:

> **WIRELESS TELEGRAPHY** – The fascination this subject possesses for the world at large has led the Marconi companies to commence the publication of a magazine call the 'Marconigraph,' the aim of which will be to set forth, month by month, all the interesting happenings incidental to the development of their marvellous system. The first number is just issued, and is an admirable production. The frontispiece is, appropriately, a fine photograph of Mr. Marconi. A rough calculation has been made recently as to the success and value of the assistance rendered by wireless telegraphy to ships at sea in the saving of life, and it is estimated that 3,000 persons owe their continued existence at the present time to the help rendered by wireless telegraphy. The magazine records that very successful tests have taken place with the latest pattern of the Cavalry Field Station. The tests were carried out between Broomfield and a position 4 miles on the other side of Ipswich, a distance of over 40 miles, and good readable signals were obtained on all three wave lengths. The appointment is also announced of Mr. Alfred Eddington, formerly assistant in the test room of the Company's works at Chelmsford, to be assistant works manager.

Dr. James C.H. Macbeth, noted cryptographer and one of the early members of the Company wrote:

> 'Godfrey Isaacs, Managing Director of the English Marconi Company, was the king-pin of the organisation. He was an extravagant promoter and he had an insatiable love of power; he was the salesman of wireless with the business strategy and enthusiasm necessary to promote such a radically new communication system. He revelled in acquiring telephone and electrical instrument companies to link them as subsidiaries of wireless. He was generally faced with litigation, and from that Marconi, who detested routine business and legal conflicts, suffered pangs. Yet he entrusted the business end of wireless and its promotion to Mr. Isaacs, who presided at the Company's meetings and usually at public functions. Speech-making and writing were sacrifices for Marconi; in either he was concise.'

Godfrey Isaacs applied himself energetically to the problem of capitalising the expansion of wireless telegraphy and started litigation all over the world to prevent further infringement of Marconi's patents. By 1912 Marconi had established companies in Russia, Spain, the Argentine, Canada and America, all using the same master patents and all associated with the British Company, which usually held the majority of the shares and placed Directors on their Boards.

Isaacs knew the Company had to grow to survive and he was determined not only to enforce the Company's patents, but also to systematically put out of business the Company's four main British competitors. In England the British Radiotelegraph and Telephone Company (BRTS) were marketing a wireless system developed by John

Graeme Balsillie, an Australian engineer and inventor.

The British Insulated and Helsby Cable Company were marketing their own wireless system and a syndicate headed by Scottish electrical engineer Alexander Muirhead and Marconi's old rival Professor Oliver Lodge also marketed their own wireless system.

But perhaps the biggest threat was the German Telefunken Company, backed by the German Government. This had been born out of an amalgamation of Germany's wireless companies and the work of many of its early pioneers, even if much of this had been appropriated from Marconi during his early experiments.

In 1911 Marconi's started legal action for patent infringement. The ruling in favour of the Marconi Company was issued in the case brought against the 'British Radiotelegraph and Telephone Company' for the infringement of patent No. 7777 of 1900. The courts decided in favour of the Marconi Company recognising the validity of his patent. Marconi also sued the Helsby Company in 1913 and won. In 1912 Isaacs also brought the exhausting 14 year legal battle with the German wireless company Telefunken to an end when the German combine applied for a licence for use of Marconi's master wireless patent.

Then Isaacs finally resolved a legal muddle left by an old quarrel between Marconi and Professor Sir Oliver Lodge that dated back to the earliest days of wireless. Isaacs persuaded him to accept **£18,000** for his patents and Company and join the Marconi Company as a scientific adviser. In doing so he also effectively closed down the competing Lodge-Muirhead Company. Isaacs had efficiently cleared the field of all major competitors, but even he could not have realised that 1912 was to be a both a year of tragedy and a major turning point in the fortunes of the Marconi Company.

As the New Year dawned, Isaacs was reasonably pleased with the progress of the Company.

He had cleared the way of competitors and it looked highly likely that the British Government would soon sign a contract for the 'Imperial Wireless system'. In March 1910 Godfrey Isaacs had submitted to the Colonial Office an imaginative plan to link the British Empire by a network of short wave wireless stations and on 7th March 1912, Mr. Samuel, the Postmaster-General, wrote a letter accepting the Marconi tender.

Although this new bold system would be delayed for a further decade the future of wireless and the Marconi Company looked bright, although they still had not returned any dividend to its shareholders despite fourteen years of world changing developments.

CHAPTER FOUR

Titanic

On 14th December 1911 the Norwegian explorer Roald Amundsen and his team became the first humans to reach the Geographic South Pole. The English explorer, Robert Falcon Scott had also returned to Antarctica with his second expedition, the Terra Nova Expedition, in a race against Amundsen to the Pole. Scott and four other men reached the South Pole on 17th January 1912, thirty-four days after Amundsen. On the return trip, Scott and his four companions all died of starvation and extreme cold and some of their bodies, together with journals and photographs were discovered by a search party eight months later.

On 10th April 1912 the new British ocean liner RMS *Titanic*, the largest passenger ship of her time left Southampton on her maiden voyage for New York City. Equipped with the finest luxuries, the massive vessel was truly a spectacle, with an on-board swimming pool, a gymnasium (complete with an electric exercise horse), squash court, Turkish bath, veranda cafe and libraries on both the first and second class decks. Each room was decorated with beautiful French polished mahogany furniture. In addition, the *Cafe Parisien* offered first class guests a dining experience unlike anything ever seen before on a cruise ship.

She was also equipped with steam-powered generators, an electrical subsystem which provided lighting to the entire ship and two Marconi wireless systems. She was the pride of the White Star Line and joined her sister ship, the RMS *Olympic* which had already enjoyed great success and acceptance by the travel industry. The *Titanic* was also considered to be 'virtually unsinkable'. On 11th April *Titanic* arrived at Queenstown, Ireland (today known as Cobh) picking up her final complement of passengers before steaming westward for New York. On 14th April the *Titanic* struck an iceberg in the northern Atlantic Ocean at 11:40 p.m. (local time as in 1912 ships at sea did not use standard time zones. Most wireless messages were logged in either GMT or New York mean time (exactly 5 hours behind GMT), depending on which side of the 40°W longitude line they were on. At noon on 14th April 1912, *Titanic*'s clocks were 2 hrs 58 min behind GMT).

At 2:20 a.m. (local or 5.47 a.m GMT) on 15th April 1912, less than three hours later, the *Titanic* sank beneath the freezing water, taking with her the lives of more than 1,500 people.

The sinking of the *Titanic* and the huge loss of life shocked the world, but the tragedy would have been much greater had not 711 people been rescued, plucked out of their life rafts in the middle of the freezing ocean by a ship summoned by wireless while she was over 58 miles away.

In 1912 communication by wireless at sea, especially between ships, was still in its infancy, even though Marconi had established a transatlantic wireless message service four years earlier. The first five years of Marconi's career had been a constant battle to develop wireless communication. He had faced massive engineering challenges, strong competitors, customer indifference, lack of funding, political and industrial espionage and rejection of his ideas by the established Victorian scientific community. But through it all Marconi had persevered and made wireless a reality.

On Wednesday 28th February 1900, the German liner *Kaiser Wilhelm der Grosse,* carrying 1,500 passengers became the world's first commercial merchant ship to be fitted with Marconi wireless equipment built at Hall Street. But by 1912 fewer than 400 ships had been equipped and the intent of most wireless installations at sea was only to profit from the transmission and receipt of messages. Thus, the *Titanic*, like other large ocean liners was equipped with Marconi wireless systems primarily for handling wireless message traffic for revenue. It was the responsibility of the wireless operators, all of whom were employed by Marconi's and who sailed with the ship and their equipment to transmit and receive messages known as 'Marconigrams'. These included personal messages, general greetings, stock exchange quotations, business communications and news services.

The use of wireless for signalling distress and emergency was incidental. The sentiment of the period was that ships not carrying passengers (unless they were part of the Royal Navy) simply did not need to be equipped with wireless equipment. Those that were usually operated a normal daily schedule in line with their passengers' demands; hence 24 hour cover or emergency watches were unknown. When the wireless operator went to bed the equipment was simply turned off.

Although Marconi had been convinced since the earliest days of his experiments that wireless was essential for safety at sea, he faced a constant battle to get wireless systems accepted. The absence of any coherent regulations governing both safety of life at sea and wireless operation undoubtedly contributed to the *Titanic* disaster and the huge loss of life. After the *Titanic* struck the iceberg, the importance of the wireless room and the ability to communicate between ships was recognised as being essential.

On the evening of 14th April 1912 the night was clear and starlit. The sea was calm. This was a rare event in the North Atlantic. These conditions were a sailor's dream and *Titanic's* Captain E.J. Smith apparently saw the opportunity to go full speed ahead in an attempt to break the ocean crossing speed record (the *Blue Riband*) held by the rival Cunard liner the RMS *Mauretania* since 1909. Smith was an extremely experienced (sea) Captain and Master for all the White Star Line's maiden voyages. After *Titanic's* first voyage he was due to retire and it has been speculated that in pursuit of the record on his last command he may well have thrown caution to the wind. This may explain why the *Titanic* apparently failed to heed ice warnings received by wireless.

At 11:40 p.m., a lookout repeatedly shouted the warning of 'iceberg dead ahead!!' The bridge attempted evasive action and full reverse propellers was attempted, but it was too late. The inertia of the *Titanic* was too great and she continued to plunge ahead on a path that ripped open the sealable bulkheads on the starboard side, a rupture estimated to be about 300 feet in length. The seemingly invincible *Titanic,* having lost its sealable airtight bulkheads was no longer able to remain buoyant at its bow. Within two and a half hours of striking the iceberg, the bow submerged. At 2:10 a.m., the massive stern rose into the air and the ship plummeted to a watery grave, carrying with it 1,513 helpless and terrified passengers.

Earlier, aboard the SS *Californian*, the Marconi wireless operator Cyril Evans turned on his wireless to clear his routine traffic. But being only ten miles from the *Titanic*, the operator on duty on the *Titanic* strongly advised Evans to 'shut up', as he was interfering with their commercial wireless traffic to the Cape Race wireless station in

Newfoundland. Evans complied. Being the lone wireless operator on the *Californian* and having worked a long day, Evans switched the equipment off and retired for the night. The *Californian,* within sight of the Titanic, had found itself in the same ice field earlier in the evening at 11:00 p.m. Wisely, Captain Arthur Rostron of the *Californian*, ordered his ship to a complete halt, intending to carefully wend his way out at daybreak.

The *Titanic* struck the iceberg at 11:40 p.m., less than a minute after the first sighting of the iceberg by the lookout. But the first 'CQD', (General Call Distress) was not initiated until 12:15 a.m., thirty-five minutes later. The *Californian's* First Officer observed white flares shot into the sky from the *Titanic.* Unfortunately, he assumed the flares to be part of a celebration aboard the *Titanic* as was the custom during heavy partying. He also considered the possibility that he was observing shooting stars. In 1912 the arbitrary discharge of white or coloured flares was acceptable and commonplace as there were no regulations governing the deployment of flares as a signalling or emergency tool. The uncertainty of the First Officer nevertheless prompted him to use a Morse code signalling lamp aimed at the Titanic, but he received no light signal response. Neither the First Officer nor the Captain of the *Californian* attempted to wake Evans and direct him to the transmitting key of the wireless to send a message of inquiry to the *Titanic*. With this one failure aboard the *Californian*, within sight of the *Titanic,* the fate of the 1,513 lives was sealed.

Fifty eight miles to the southeast of the *Titanic* was the SS *Carpathia*. The ship's wireless operator Thomas Cottam was preparing to retire when by chance he initiated contact with the *Titanic* to advise its operator that the Marconi land station at Cape Cod was attempting to contact him. The response from the *Titanic* was prompt, with an urgent message distress message and requesting immediate aid. The *Carpathia* turned its course 140 degrees and headed for the *Titanic* at full speed. Altogether, eight ships over a wide area heard the *Titanic's* 'CQD' distress call and were racing to the scene, including the *Frankfurt* from over 140 miles away.

The *Carpathia* was the first to arrive at the scene of the disaster at 4:15 a.m. There was no *Titanic*, only an empty sea dotted with lifeboats, adrift without lights and the shivering passengers in them huddled altogether to protect against the freezing air. By 8:30 a.m. all survivors had been picked up, the *Carpathia* had recovered 14 lifeboats and 712 survivors, although one of them died later en route to New York.

By then, the world was waking up to the news that the unsinkable ship had been lost. Families with loved ones aboard the vessel were desperate for a list of survivors, but it wouldn't be compiled until a week after the accident. One can only imagine the despair of the *Californian* crew in the morning when they were told by wireless of the *Titanic's* loss. They were only a few miles away but they were the last to know when the wireless operator returned to his station in the morning to begin routine traffic operations.

The man who gave the world wireless had also nearly been lost in the *Titanic* disaster. Marconi and his family were invited by the White Star line's Chairman, Bruce Ismay, to be his guests on the maiden voyage of the *Titanic*. Fortunately for Marconi he had to hurry to America on business, so he went earlier on the RMS *Lusitania*. Travelling with the Managing Director Godfrey C. Isaacs and his son Marcel, Marconi had arrived in New York on 15th March. They had proceeded to the United States in connection with the affairs of the American Marconi Company, which was controlled by the English company. In addition, as part of Isaacs world-wide purge Marconi's had brought a legal action against the United Wireless Company for patent infringement that was being heard in the Federal Court on 25th March. The Company was in liquidation and with some of its directors in prison the English Marconi Company purchased the assets of this company and resold them to the American Marconi for 1,488,800 fully-paid shares of $5 (£1) each in the latter company.

Marconi's wife Beatrice retained her booking on the *Titanic*, but on the eve of the voyage, she too cancelled due to their son Giulio's sudden illness with a high fever. She cabled Marconi that she had to postpone her trip and the *Titanic* sailed on her fateful voyage on 10th April without the family aboard. Marconi still held a ticket for the *Titanic's* return voyage to England, planned for 20th April 1912.

Marconi arrived in New York just in time to hear that a wireless message had been received at the Cape Race wireless station in Newfoundland which might indicate a disaster at sea. The *New York Times* promptly sent a wireless message to the *Titanic's* Captain, Edward J. Smith. They got no reply.

A period of total confusion ensued. The full horror and tragedy of the disaster was only fully comprehended when the *Carpathia* sailed into New York's Harbour through the rain on Thursday night. As soon as her gangplank went down, Guglielmo Marconi stepped out of the immense and silent crowd at Pier 54. With police clearing the way, he was one of the first people along with Mr Speers of *The New York Times* to go aboard to interview the wireless men whose work had saved so many. The *Carpathia's* wireless operator Thomas Cottam and *Titanic's* second wireless officer, Harold Bride were onboard. The *Titanic's* first wireless officer, John George Phillips, had been lost in the disaster.

Suddenly wireless was the hottest news topic in the world and the New York Electrical Society invited Marconi as a guest lecturer on 17th April 1912, just days after the disaster. The loss of the *Titanic* was the main headline in all newspapers. Marconi, praised as the saviour of 711 lives was in the forefront of the news as never before. It seemed that all New York wanted to see him before he returned to Europe, and the Engineering Societies auditorium was jammed to capacity. When the inventor appeared at the side of the platform, the crowd in the balcony saw him first, and the cheering began. It spread to the main floor and was continuous as Marconi bowed many times.

On 18th June 1912, Marconi started to give evidence to the Court of Inquiry into the loss of the *Titanic* regarding the marine telegraphy's functions and the procedures for emergencies at sea. The principal finding was inescapable. Without wireless on board the *Titanic*, all 2,224 passengers and crew would have perished.

The two Marconi wireless operators on board and their Hall Street manufactured equipment undoubtedly saved the lives of over 730 people. John George 'Jack' Phillips served as the senior wireless operator and Harold Sydney Bride was the junior wireless officer on board. As the *Titanic* was sinking they worked tirelessly to send wireless messages to other ships to enlist their assistance with the rescue of the *Titanic's* passengers and crew and both men remained at their posts until the ship's power failed. Phillips managed to find an overturned lifeboat to cling to, but he perished in the icy Atlantic Ocean. Bride was washed off the ship as the boat deck flooded, but managed to scramble onto the upturned lifeboat Collapsible 'B', and was later rescued by the *Carpathia* the following morning. Despite being injured, he helped the *Carpathia's* wireless operator transmit survivor lists and personal messages from the ship.

The last picture of the RMS *Titanic*

RMS *Titanic* wireless room

Mr Punch honours Mr Marconi

RMS *Olympic* wireless room

The popular press of the day hailed Marconi as a hero. A tribute to Marconi printed in *Punch* Magazine after the *Titanic* disaster saw the cartoon character Mr Punch doffing his hat to the Italian pioneer, with the following caption:

'Many Hearts Bless You Today Sir, The Worlds Debt To You Grows Fast.

As Lord Samuel, the British Postmaster General at the time, stated: 'Those who have been saved have been saved through one man, Mr Marconi and his wonderful invention.'

A few days later the survivors of the *Titanic* presented Marconi with a solid gold medal, in gratitude for Marconi's wireless installation on board the *Titanic* credited for saving their lives. The survivor's cries of *'Ti dobbiamo la vita!'* [*We owe you life* !], remained with Marconi for the rest of his life. Four weeks after the *Titanic* disaster honours and editorial tributes were still being heaped upon Marconi, including on 21st May 1912 the award of The Grand Cross of the Order of Alfonso XII from Spain.

It is possible that the *Titanic* disaster could have been averted had the lessons been learnt from the earlier *Republic* accident at sea in 1909. Following that collision, Jack Binns the Marconi wireless operator on board the *Republic*, had immediately started to send 'CQD' distress signals and many ships rushed to its aid. For two days in freezing conditions Jack Binns sent out a total of two hundred messages to help guide rescuing ships to his stricken vessel's position. All 461 passengers, except for five crew members survived the disaster and were transferred to the S.S. *Florida.* The SS *Republic* was abandoned and the SS *Florida* herself in danger of sinking transferred all passengers to the S.S. *Baltic* for the return trip to New York. The combined number of passengers of the two ships *Republic* and the *Florida* totalled 1,650. Although hailed as a great triumph for wireless communication, the small loss of life meant that there was no public outcry following the near disaster. Thanks to wireless all the passengers and crew who had not been killed by the initial impact were rescued.

But it was only after the *Titanic* went down, with huge loss of life that the ensuing inquiries both in Great Britain and the United States led to far reaching legislation.

The most significant result of the disaster investigations was the call for an International Radio-Telegraphic Conference, to convene in London, on 5th July 1912, for the purpose of establishing regulations and procedures governing wireless services aboard ships and shore stations. Attended by sixty-five countries, regulations and procedures were enacted, some of which are still in effect today. Among these is the use of 'SOS' as the universal call of distress as it was determined to be the simplest form of signalling to replace 'CQD'. The abbreviated 'Q' code signals that are still in use today were also an outcome of the meeting. There was also agreement on a common wavelength for ships' wireless distress signals. Also, every ship was instructed to maintain wireless silence at regular intervals, when operators should listen for distress calls.

The 'Safety of Life at Sea' Conference was held in London on 12th November 1913 and was also attended by sixty-five countries. This conference was the turning point for wireless communication at sea. Sweeping regulations were put into effect governing all ships at sea, whether motor propelled or under sail. Ocean going vessels carrying passengers to foreign ports were mandated to be fitted with a wireless communication system. Further, the ship's wireless room and shore stations were to be manned twenty-four hours a day. Now the wireless room became the focal point on board with all vessels having to abide by all new rules, regulations and laws, establishing safety of the passengers and ship as the first priority. The value of wireless on board was now self-evident. The 'Safety of Life at Sea' Conference concluded on 20th January 1914. It was determined that all countries having ocean going vessels carrying passengers were culpable for inadequate safety regulations on its vessels. The Conference emphasised the necessity for united action to revise the old laws and adapt them to new conditions.

On 9th October 1913, Wireless was again instrumental in the saving of 650 lives from the SS *Volturno,* an ocean liner that burned and sank in the North Atlantic. She was a Royal Line ship under charter to the Uranium Line at the time of her fire. At about 06:00 a.m. the *Volturno,* carrying mostly immigrants bound for New York, caught fire in the middle of a gale in the North Atlantic. The crew attempted to fight the fire for about two hours, but, realizing the severity of the fire and the limited options for dousing it in the high seas, Captain Francis Inch had his wireless operator send out SOS signals. Eleven ships heeded the calls and headed to *Volturno's* reported position, arriving throughout the day and into the next. Wireless had done its job, but tragically several of *Volturno's* lifeboats with women and children aboard were launched in heavy seas, all the boats either capsized or were smashed by the hull of the heaving ship, leaving no one alive.

This near disaster and the tragic sinking of the *Titanic* in 1912 focused public attention on the importance of ships being equipped with wireless and the necessity for maintaining a 24 hour communications watch. After the loss of the *Titanic* many countries required that ships over 1,600 tons be equipped with wireless equipment. This caused demand for marine equipment to soar. Within a decade, over 500 shore stations had been built, establishing wireless (now almost universally known as radio) communication worldwide.

The Marconi Company share price that had been languishing at an unsaleable ten shillings or less, rocketed almost overnight to £10.00 and in 1913 the Marconi Company was finally able to declare its first dividend to its shareholders. It had taken 16 years of intense struggle.

It was the tragedy of the *Titanic* that gave birth to the modern wireless age and spurred a growth in manufacture and development that probably would not have occurred otherwise. What better advertisement for the 'wireless' could there be as the news media printed the disaster story day after day for months? The mystique and the magic behind the word 'wireless' also gave birth to a new generation of aspiring operators and engineers, together with the need for accelerated manufacture of wireless equipment to fulfil the demands of ship and shore installations.

The tragedy of the *Titanic*, occurring when it did during a period of slow growth in a new industry was responsible for the kick start of the wireless, radio and electronics industry that today still provides the greatest number of jobs in the history of civilisation.

The Marconi Company recognised the need for operator training and established Marconi wireless training schools throughout the world, including the major cities of the United States. The new regulations requiring wireless on board all ocean going vessels made it necessary for the Marconi Company to significantly step up equipment production to meet this need. To do this the Company urgently needed new manufacturing facilities on a scale never before imagined in the still embryonic industry.

All this was the legacy of the *Titanic* disaster.

Made in Hall Street

Poldhu Point Transatlantic Station

Opened in 1901 Poldhu was the largest station in the world. It transmitted the famous Morse code letter S (---) across the Atlantic. The original aerial mast is shown above and the temporary structure after the storm. The Poldhu operating bay above and the crucial spark gap (left) were all made at Hall Street.

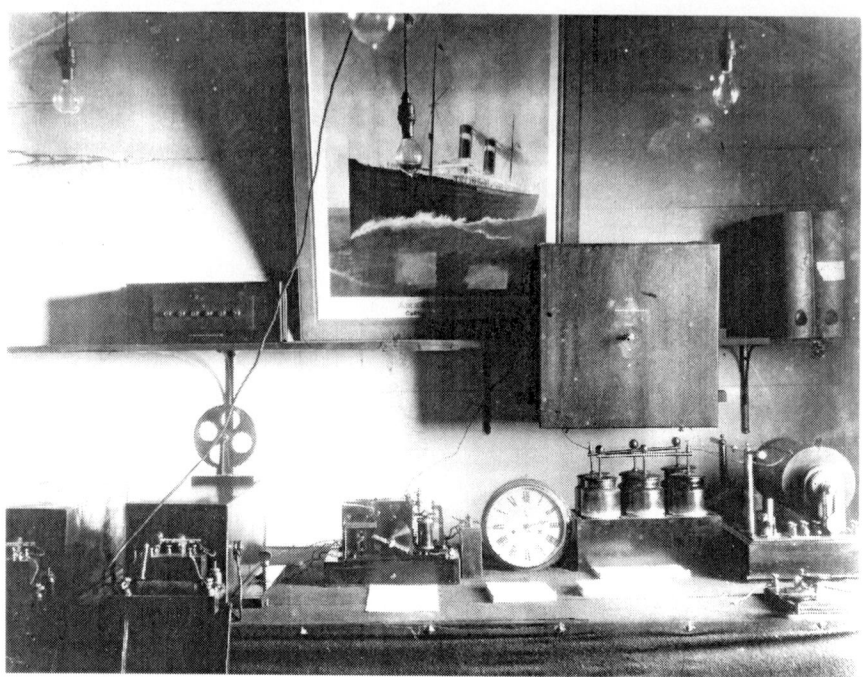

The Lizard receiving station crucial for Marconi's transatlantic experiment - all built at Hall Street - proved conclusively that wireless signals were not obeying a 'line of sight rule' when the Crookhaven station was received over 225 miles

Marconi at the Cabot Tower having crossed the Atlantic - the equipment in front of him was part built at Hall Street

SS *Philadelphia* wireless cabin - all built at Hall Street - whose voyage in February 1902 proved conclusively that long distance wireless communication was a fact

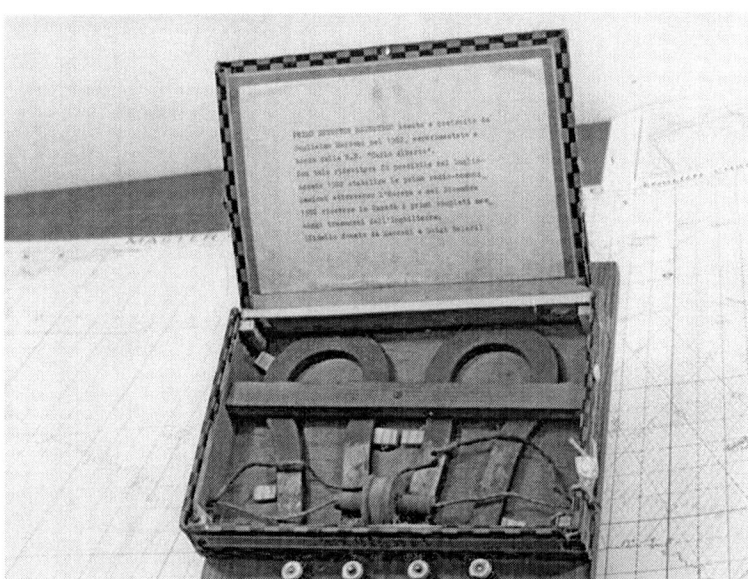

The Marconi *'Maggie'* -

Magnetic Detector that moved wireless equipment reception to the next generation. This equipment was manufactured at Hall Street continually until the factory shut its doors in 1912

Left - Marconi *'Jigger'* the crucial invention that allowed Marconi to tune both receiver and transmitter - one of the main products made by Hall Street

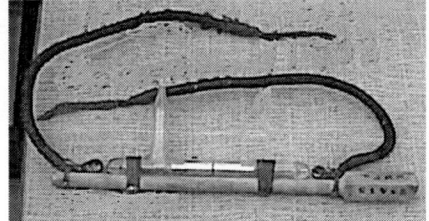

Coherer detector - the core of any wireless receiver until the advent of the Magnetic Detector above

When the New Street factory opened the ITU delegates were shown horse mounted wireless equipment destined for the army - manufactured at Hall Street.

Above the main transmitter, below the generator and operating bay.

At the start of the First World War the British army, unlike the Royal Navy, were unconvinced about the use of wireless in warfare.

A Marconi-Bellini-Tosi is a type of radio direction finder (RDF), which determines the direction to, or bearing, of a radio transmitter.

Invented by a pair of Italian officers in the early 1900s, they joined forces with the Marconi Company in 1912.

THE OPERATING CART.

Marconi Hall Street manufactured wagon wireless equipment -

2 kW wireless stations with the upper unit destined for the Serbian army in the First World War

.

CART "B" FOR PORTABLE SET

REFERENCE
A	ALTERNATOR
B	ENGINE
C	BASE FOR ENGINE AND ALTERNATOR
D	BOX CONTAINING MAST FITTINGS
E	PETROL CAN
F	TENT IN CANVAS BAG

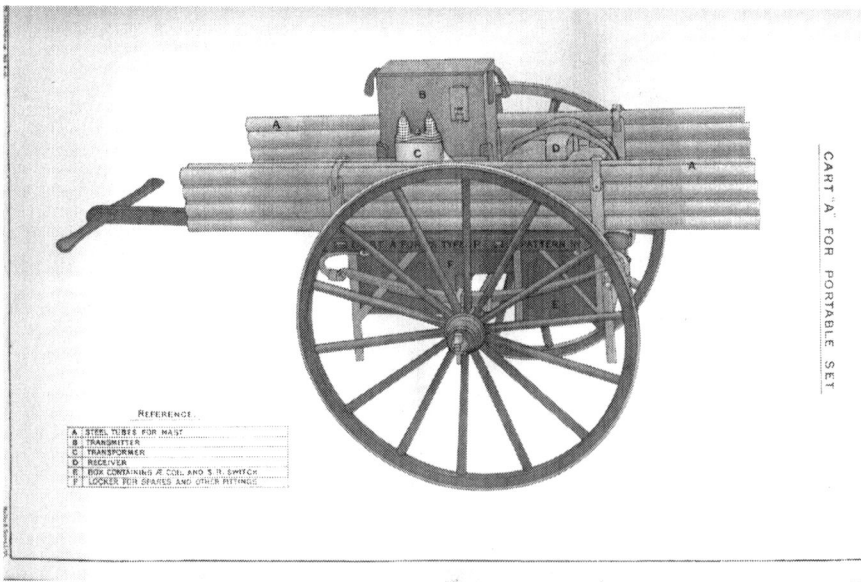

CART "A" FOR PORTABLE SET

REFERENCE
A	STEEL TUBES FOR MAST
B	TRANSMITTER
C	TRANSFORMER
D	RECEIVER
E	BOX CONTAINING Æ COIL AND S.R. SWITCH
F	LOCKER FOR SPARES AND OTHER FITTINGS

Hall Street manufactured army wagon wireless equipment -
The lower wagon held the portable sectioned wireless mast

CHAPTER FIVE

NEW STREET

The World's First Purpose Built Wireless Factory

Even before the *Titanic* set sail, plans for expansion within Marconi's were already underway. It was clear to Isaacs that the original works at Hall Street was becoming too small as the demand for the new wireless telegraphy equipment had already increased tenfold.

Again the Marconi Company struggled for space. There was insufficient space in Broomfield for any form of major expansion and no other current site offered a viable alternative. Having learnt from their mistakes with the ill-advised and expensive move to the Dalston factory in 1905, in January 1912 Isaacs proposed building the world's first-ever *purpose-designed* and *purpose-built* wireless factory on the local cricket ground in Chelmsford that was owned by the Church Commissioners. The proposed new works would cover the whole of the site. To the north two new roads would be constructed leading from New Street parallel to Rectory Lane (Marconi Road and Bishop Road) where cottages would be built for the Company employees. The site would be known as the Marconi New Street works.

The Marconi Company commissioned the architects William Dunn and Robert Watson in London to draw up plans for the first factory to be specifically designed for the construction of Marconi's wireless equipment. Initially the architectural practice operated from 35 Lincoln's Inn Fields and by temperament, Dunn was more a mathematician and structural engineer than an architect and he had a particular flair for design in concrete, while the decorative aspects of the partnership's work fell to Watson.

The Managing Director, Godfrey Isaacs, wanted the new factory to be finished and working by mid-June 1912, an almost impossible target, but he wished to show off his smart new wireless factory on 22nd June to his leading competitors, Government officials and other experts who would then be in London for the International Radio-Telegraphic Conference, due to start on 5th July 1912. This was the event instigated as a direct consequence of the *Titanic* disaster and resultant enquiry.

Godfrey Isaacs's plans were in fact not just for a new factory; he wanted the new complex to be a complete self-contained village within a town. An agreement was made to purchase the local Cricket Ground. Chelmsford Cricket Club could trace its history back to 1811 and in 1879 the Club had moved to the land adjacent to New Street which became their home for the next 32 years. But the club now became a victim of Chelmsford's rapid industrial growth.

As the area was Glebe land, adjacent to the Rectory on Rectory Lane and belonged to the Church, it had to be purchased from the Bishop of St Albans, the Rector of St Marys and the Ecclesiastical Commissioners. Once the contract was signed the cricket ground was pegged out on 10th February 1912, and the bricklaying, using 500 men, started on 26th February. Despite a short building strike, just seventeen weeks later the changeover from Hall Street to the new 70,000 square-foot (6,500 sq m) factory complex was accomplished in just one weekend, leaving the old silk mill (of circa 4,000 square-foot) once more empty and abandoned, although the wireless station remained operational. Amazingly all sections of the new works were fully functioning in time for the International Radio-Telegraphic conference.

The site was ideal for Marconi's plans. Located due north from Hall Street, the ex-Chelmsford Cricket ground had an area of around 10 acres. The large railway coal yard opposite gave easy access to the main London Great Eastern Railway. Rail trucks could bring coke to feed the Company's power station, and the heavy components could be loaded directly onto trucks for nationwide delivery or to the docks for shipment overseas.

No expense was spared in the works design; arches were prominent, both internally on office doors and on the south side of the building facing the courtyard. The new works were designed from the outset for mass production and equipped with the latest and best tools, apparatus, test rooms and laboratories. The factory frontage ran alongside New Street with a clock tower above the main entrance. This building housed the Company offices, drawing office, and showrooms. The workers' entrances, (men and women had separate doors) were to the left while a separate building contained the separate male and female dining areas and clubrooms. As at Hall Street typically men worked on machine tools in the machine shop or in carpenters' shops, whilst the women worked producing induction coils for the spark transmitters and other more delicate tasks.

Behind the office block was the main test area, above which six glass cupolas provided light. Behind this was the factory where overhead line shafts were driven by DC motors, supplied in turn by state of the art steam turbine driven DC generators in the power house. In the courtyard a single 200 foot wireless mast was erected to support an antenna for transmitter testing.

During construction the building required 2.5 million bricks, 400 tons of steel and 9,000 truckloads of earth were removed. Sewers were diverted and a 400 foot well was sunk. The completed factory frontage was 200 feet long and 40 feet wide and the whole site was fitted with low pressure hot water radiators throughout. Railway lines were extended across New Street from the main shunting yard and the national network beyond and new sidings were constructed on one face of the works with two tracks, one in and one out with a drive way running between. The packing department had two loading bays with turntables and two electric capstans to winch the carriages in and out of the works so they could be loaded and unloaded under cover. The factory building arches were built to accommodate the exact size of the railways carriages so they could be manoeuvred in and out for loading. At the entrance to the rail sidings there were two weighbridges checking the weight of raw materials delivered. The factory was designed to be self-sufficient whereever possible and to manufacture all its components and sub-assemblies in house. Hence wood and iron, steel and packing materials would continually arrive and completed equipment depart via the railway.

The main New Street Works building behind the New Street frontage was 466 feet by 150 wide with its roof constructed to a 30 foot saw tooth spacing design with 65 wooden shafts to give ventilation. It was capped with green slates on felt and match boards. The whole glazing system used was Rendle's 'Invincible' glazing, normally used in railway stations and large public buildings. The system used steel T bars with specially shaped copper water and condensation channels, all formed in the one piece and resting on top of the T steel; the glass rested on the zinc channel and a copper capping was fixed over the edges of the glass and secured with bolts and nuts. The floors were tongue and groove Ash wood blocks laid over the concrete floor, while the walls were Fletton bricks. The Fletton Brick Company eventually became the London Brick Company and the dominance of London Brick in the market during this period gave rise to some of the country's most well-known landmarks, almost all built using the ubiquitous Fletton.

The power test area had granolithic floors due to the heavy machinery involved along with a five ton, three motor overhead travelling crane on a runway that extended over the loading dock in the packing department. Fire risk precautions were taken very seriously in the new factory. Fireproof doors separated each department and the entire building had been installed with automatic glass disc Grennell sprinklers. Although commonplace today, sprinklers were a new safety innovation in 1912 but considered essential in a factory where spark transmitters, high power electricity, generators, wood working and packing materials all existed practically side by side.

The builders selected for the project were Cubitt & Co Ltd (actually Holland, Hannen & Cubitt's Ltd) a major building firm based in London who were responsible for many of the great buildings of London. The Company had been formed from the fusion of two well-established building houses that had competed throughout the later decades of the nineteenth century but came together in 1883 when Holland & Hannen acquired Cubitts, a business founded by Thomas Cubitt some 70 years before. The combined business went on to construct many important buildings and structures including the Prudential Assurance building in High Holborn (1906), the Cunard building in Liverpool (1917), the Cenotaph in London (1920), London County Hall (1922) and South Africa House in London, completed in 1933.

The New Street project was so successful that the Company awarded the same building company construction of its new long-wave transatlantic transmitting station near Waunfawr, on the slopes of Cefn Du, three miles east of Caernarfon. The first land was purchased in December 1912 and work started on the 21st February 1913, the main building measuring 100 feet by 50 feet and elevated 830 feet above sea level. In 1915 the Admiralty required a powerful spark transmitter to be built at Moody Brook to the west of the Stanley settlement in the Falkland Islands. The contractors to the Admiralty were Marconi's who again sub-contracted construction to Cubitts who employed 20 labourers and duly despatched them to the Falklands.

The 250 International Radio-Telegraphic Conference delegates made their inspection tour of the brand new Marconi New Street Works on 22nd June 1912. A special train brought delegates to Chelmsford from London for the grand opening of the new factory. The Mayor of Chelmsford, Alderman T.J.D. Cramphorn, J.P, accompanied by his sister, the Mayoress, stood in the new entrance hall to welcome Marconi, his team and the world's leading wireless experts. It was a great honour to have all the delegates gather in his Borough, including representatives from as far away as Egypt, Japan, Turkey, Morocco and Siam.

The works tour included a full working demonstration of the new Marconi Wireless Telegraphy system. From the main hall over the polished wood floors the visitors were led to the left past the offices, where they were taken outside to the far side of the factory and the factory tour began at the carpenters' shop where woods such as mahogany and teak were being crafted into mountings for the wireless components. The visitors were then taken into the machine shop, at the time one of the finest and largest in Essex, 187 feet. long by 90 feet. wide. Here large D.C. motors powered two overhead line shafts driving the individual machines. From here the visitors continued through raw stores, where tons of ebonite and brass were stored. The next stop was at a point below the water tower that housed a large 8,000 gallon tank connected to the Grinnell fire sprinkler system. The tank was filled from the 400 foot borehole dug into the chalk that was lined with 6 inch steel tubes to a depth of 333 feet.

Upper:-   Marconi New Street works Ground Breaking Ceremony
Lower:-   Foundations Underway

Marconi New Street works
construction.......
in just 17 weeks

The New Street build team

New Street Works

Visit by International Radio-Telegraphic Conference delegates

Marconi personally hosted the New Street Works visit by the International Radio-Telegraphic Conference

New Street works main workshop using much of the equipment movded from Hall Street

New Street Works, c. 1920

The two massive 450 foot high aerials dominated the New Street works and were used for the early telephony broadcasts in 1920 that gave birth to British Broadcasting

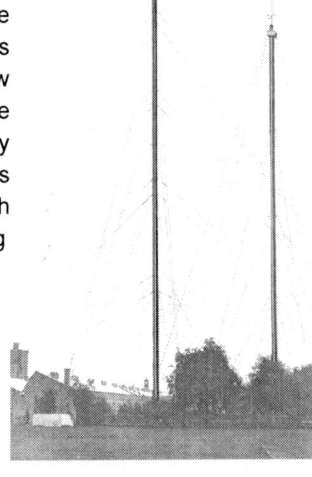

On their whirlwind tour of the new works the visitors were next taken outside to the railway sidings where railway trucks were delivering the raw materials and despatching completed wireless telegraphy equipment. Opposite the railway siding could be seen the powerhouse containing a number of steam turbines including a 2,780 r.p.m., 45 hp turbine. From here the delegates were taken through the packing department where the completed wireless equipment was packed for shipping; Finished Stores for items awaiting packing; and finished parts stores, where sub-assemblies were housed; to the condenser and winding shop, the winding shop being staffed entirely by women.

The tour continued on to the mounting shop where wireless telegraphy sets could be seen on an assembly line, being mounted onto army carts. One of the problems during the Boer War in South Africa had been the cutting and tapping of the telegraph wires by enemy raiding parties. The new Marconi Wireless Telegraphy system of course overcame the problem and many of the carts shown to the delegates had been built at Hall Street and were already in use in Tripoli by the Italian Army. From the mounting shop the delegates were ushered to the test room, a long shop, situated behind the main façade alongside New Street, daylight entered the room from the six cupola skylights along its length. On occasion the discharge from a spark gap transmitter on test sounded like a rifle shot and made each group on the delegate's tour jump with much nervous laughter.

The tour still continued via the power and oil test departments back to the main building and the showroom where many examples of the Marconi wireless telegraphy were on show. During the tour, visitors could see a replica of a ship's wireless cabin; military equipment including cavalry wireless apparatus, carried by horses, various valve receivers, a 3 kW set with quick-change tuning of the primary circuits, a 5 kW battleship set, a 1½ kW ship set, a new ½ kW cargo set, and also a wireless equipped car, where an operator communicated with Hendon Airport. As the New Street works had hardly started operations yet, much of this equipment had been brought across the town from the Hall Street works.

The Company had spared no expense to make the tour the showcase for all that the new technology and the new factory could offer. Marconi engineers had installed the new Bellini-Tosi direction finding system, by which the direction of a transmitting station relative to the stations receiving the message could be ascertained. The experimental system had only that year first been installed for trials aboard the RMS *Mauretania*. The highlight of the tour was a demonstration of transmission and reception between the new works and the Wireless Telegraph site at Poldhu in Cornwall. For this a 15 kW ship set was connected to the main aerial suspended from a 200 foot high temporary tubular mast outside. Transmission wavelengths ranged from 600 to 2,800 metres, the spark set emitting an almost 'musical' hissing note of 400 Hz. After touring the works the visitors were taken past the temporary mast and on to the old cricket pavilion where a Morse sender and Morse inker tape machine had been installed. The following message was sent to the Poldhu Point station:

'The President and delegates of the International Radio Telegraphic Conference present to the staff of the station their very cordial greetings.'

Poldhu replied: 'To the President of the International Telegraphic Conference, - The engineers and staff on the Poldhu station have the honour to present their respectful homage to all the delegates.'

The inspection ended in a large marquee, where the caterers Messrs. Hicks, Son and Co. provided an elegant tea, with ices, strawberries and cream. Marconi himself oversaw the whole proceedings, dressed in a dapper blue flannel suit with a fine white-stripe and a straw hat. In the evening Marconi and his Directors entertained four hundred guests to a magnificent banquet at the Savoy Hotel in London, a stone's throw from the Company's impressive new London headquarters, which had recently, and rapidly, been converted from an unoccupied apartment block in the Strand and renamed Marconi House. At the evening function the Mayoress of Chelmsford wore black satin, trimmed with pink chiffon. She and the rest of the guests were received by Marconi and Godfrey Isaacs and the party then proceeded to the impressive dining room of white and silver, adorned with flowers and illuminated with concealed lights. From the gallery an orchestra played music from various nations.

After the meal the Marconi Company presented gifts to all the delegates, each lady receiving a silver perfume scent bottle, while each gentleman was presented with a silver cigar lighter, this firing a spark on to a wick charged with petrol. The lighter was in the shape of Marconi's famous disc discharger system that he had patented in 1907. The after dinner speeches were in French, but then Marconi spoke in English, stating that for the first time since the invention of wireless telegraphy, representatives of every country in the world had assembled first in Chelmsford and then in London to help to form International regulations governing the application of this discovery. Marconi said that this was for him, personally, a great honour and he had been pleased to receive them at the Chelmsford Works. He raised his glass to those who had assisted him thanking all present, and proposed a toast to the health of the delegates.The following weekend, the delegates visited the Poldhu station and the tour concluded with a garden party at 'Eaglehurst', Marconi's private residence on the Solent.

The conference was a huge success and the New Street factory soon became a busy and vibrant manufacturing plant. The *Titanic* had sunk in April while the factory was being finished. The new International regulations that followed, bringing some good out of the disaster, meant that large volumes of new orders and work poured into the Marconi factory. Under this new load even the huge New Street site started to get congested and Marconi's started to use army huts to house various new developments. These could be seen around the factory building, in the courtyard and to house the transmitter at the base of the antenna feed. One 'smelly' problem was quickly addressed when soon after the factory opened sprinklers were added to the work's pond to prevent stagnation.

The *Essex Chronicle* in 1912 reported that the temporary 200 foot mast at New St was to be replaced by two '500 ft' masts 'as soon as the castings are completed'. In October 1912 it reported that the masts were 'rapidly rising' and they were completed by May 1913. The July 1935 report on their being dismantled also stated that: 'They were erected in 1913'.

The huge new 450 foot high masts, set 750 feet apart at the base were known locally as 'the drainpipes'. Each mast was three feet in diameter and made from four sections of half inch pressed steel plate with vertical flanges bolted together. A single flange at each end joined each section together. Five sets of insulated stays with a radius of 220 feet connected each mast with four steel anchors set into 100 ton concrete blocks while at the base of each mast a huge concrete block of 120 tons supported the load. These huge steel tubes dominated the Company's New Street Works and quickly became unforgettable landmarks, visible across the town and whole area. There was a story that Frederick Post, the mast erector Foreman had climbed the mast using just the bolt heads to enable a replacement for a broken timber top section to be hauled up during WW1.

The first two years after the completion of the Marconi factory at New Street saw the Company go from strength to strength

But the world was about to be torn apart. At 5 a.m. on 30th July 1914, with the great naval review at Spithead just over, the 'first fleet', the British Royal Navy had just left Portland. It was urgently recalled by wireless telegraphy, and instructed not to disperse for manoeuvres as had been previously arranged. The wireless signals from the British Admiralty, sent via Marconi transmitting stations moved the Royal Navy's Grand Fleet to immediate battle stations throughout the world.

On 1st August 1914 the use of wireless was forbidden to all non-British ships sailing in territorial waters. On the following Sunday 2nd August the *London Gazette* issued a special notice that it had become:

> 'Expedient for the public service that His Majesty's Government should have control over the transmission of messages by wireless telegraphy.'

On 3rd August the Admiralty prohibited the use of wireless telegraphy on all merchant ships in territorial waters, providing for the dismantling of all wireless apparatus on merchant vessels in the territorial waters of the United Kingdom and Channel Islands. On the same day a second order decreed the immediate closure of all experimental wireless telegraphy stations in this country and arrangements were made for the equipment to be impounded. The communiqué asked for the co-operation of the public in order to secure: 'Information of any wireless station which may be observed to be kept up in contravention of his orders.'

Wireless communication was now at war; a vital asset that had to be safeguarded at all costs.

Across the North Sea, the German transmitter station at Nauen also sent out an ominous call to all German merchant shipping on the high seas to make for the nearest German ports, or, if too far away, for a neutral port.

On 2nd August 1914 German troops had entered France. On 4th August Belgium was invaded. At 11 p.m. the British ultimatum to Germany expired and the two countries were therefore, automatically, at war. Urgent wireless messages were sent to all units of the British Grand Fleet: 'Commence hostilities against Germany'.

The world was now at war.

The Marconi transmitter at Poldhu Point broadcast the declaration of war across the world. The First World War bought huge and immediate changes for Marconi's. As the world's leader in wireless communication it was inevitable that the whole organisation would be turned over to war work, including production, installation, training, research and development.

Immediately on the outbreak of war the Marconi Works at Chelmsford was taken over by the Admiralty. The Clifden to Glace Bay transatlantic circuit was allowed to continue its function as a commercial station, but with interruptions and a change of wavelength to handle all naval traffic. The control of the wireless stations located at Caernarvon and Towyn passed into the hands of the General Post Office and later to the Admiralty, with Marconi's operating them for the Government. Stations for the interception of German wireless transmissions were hurriedly pressed into service at the Hall Street experimental station in Chelmsford, while the New Street factory itself was put under huge pressure to meet the demands of the armed services.

The Company's research and development departments went into overdrive and constant improvements were introduced. The delicate mechanisms employed in wireless telegraphy equipment had to be manufactured at a speed and in such volumes as had never before been contemplated. In addition, there was a constant demand for simpler instruments for the instruction of wireless operations and students, including Morse keys, buzzers, telephones and headphones.

At the outbreak of war the Royal Navy was desperately short of trained wireless operators. As merchant ships reached port, the civilian Marconi wireless operators were taken off and transferred to the Royal Navy. But this, while providing experienced men for the Fleet, in turn created a shortage in the Merchant Navy. The deficit was made all the more acute by the need to provide a much greater number of ships with wireless apparatus, as until 1914 only ships of more than 1,600 tons carried wireless and these for the most part had only one operator. Not only were the big liners deprived of their Marconi operators, but ships between 1,600 to 3,000 tons which hitherto had not been fitted with wireless now found that it was a necessity. In addition, whereas before the war it was sufficient for one operator to be carried, now it was essential that there should be at least two operators to keep a continual listening watch.

Knowing that trained wireless operators would be in great demand, for some time Marconi's had been stimulating the interest of wireless amateurs by offering prizes for competitors in Morse code examinations and by making Morse practice sets readily available.

Immediately on the outbreak of war, the Admiralty took steps to secure the services of Marconi operators for all branches of the Service. In answer to the Government's call there came an army of lads and young men from all classes who had gone straight from school to the Marconi Marine Company, in whose offices they had been trained in Morse and the operation and maintenance of wireless apparatus. These at once volunteered their services to the Admiralty and the War Office, their places being taken by other lads clamouring to be trained as wireless operators. In due course the Company provided not only an army of operators, but also technical experts whose knowledge was unrivalled.

The Company's offices were open day and night, enrolling new recruits, instructing them on the art of wireless and examining them in Morse code. At the start of the war the Company undertook to find a further 2,000 operators to augment the 3,000 already serving on merchant ships. Purpose built classrooms at King's College and Birkbeck College were made available to ease the overload of trainees from Marconi House in London.

So great was the demand, that some of the pupils and enrolled scholars were as young as sixteen. The staff at Marconi House worked to the limits of their power and to the last ounce of their energy to meet the great emergency. One thing was very certain, wireless was no longer the experimental toy of the Boer War; it was now a vital and indispensable tool for modern warfare.

In the early days of the conflict, a new wireless based science known as wireless direction-finding had been developed. This new direction-finding (D/F) equipment used a 'soft' 'C' type thermionic valve, and was a modified version of an earlier Bellini-Tosi designed directional system. It had been developed just before the war by H.J. Round, a senior engineer with Marconi's working at New Street and Broomfield.

At the outbreak of war H.J. Round's work had quickly come to the notice of the War Office, and he was soon seconded to Military Intelligence and was ordered to provide an initial two D/F stations for service in France. This was speedily done and following their success, a large network, covering the entire Western Front, soon evolved to locate enemy positions on the ground.

These stations proved to be so successful that he was instructed by the Admiralty to set up a second chain of stations in England, with the object of obtaining bearings on transmissions from enemy submarines. It was not long before similar networks were being built to maintain watch, not only for submarines but also for Zeppelins and German surface naval vessels. By 1916 the coastlines of Britain were covered by networks of Direction Finding wireless stations. Naval vessels were also fitted experimentally with D/F equipment that now included another one of H.J Round's inventions; a sophisticated error corrector.

In May 1916 the stations were monitoring transmissions from the German Navy that had been at anchor at Wilhelmshaven. On 30th May they reported a 1.5 degree change in the direction of the signals being picked up from the German fleet along with an increase in activity. The information was reported to the Admiralty who reasoned that the German fleet had put to sea. Accordingly the Admiralty ordered the British Fleet to put to sea to intercept the Germans, and the following day the Battle of Jutland was fought. It was the largest sea battle of all time. In it the British fleet lost seven ships and about 6,000 men, whilst the Germans only lost three ships (several others were seriously damaged) and around 2,500 men. While the British suffered greater losses, the battle of Jutland is considered a strategic victory for the British. While the British had not destroyed the German fleet and had lost more ships than their enemy, the Germans had retreated to harbour and at the end of the battle the British were in command of the area. Apart from two small and abortive operations the German High Seas Fleet was unwilling to risk another encounter with the British fleet and confined its activities to the Baltic Sea for the remainder of the war. Jutland thus ended the German challenge to British naval supremacy. For all his services during the war, Round was awarded the Military Cross, but as a non-combatant he refused to accept it in uniform.

As the First World War entered its second year equipment manufactured at Hall Street was to play another vital life saving role at sea even though by then the factory had been closed for almost three years.

The RMS *Lusitania* had left New York City on 1st May 1915, bound for Liverpool on her 202nd voyage across the Atlantic. Known as the 'Greyhound of the Seas,' the *Lusitania* was the fastest liner afloat, but on the afternoon of 7th May she was steaming off the coast of Ireland at less than full speed because of fog. Nor was the ship taking an evasive zigzag course.

The huge ship was a sitting duck and was headed straight into the sights of the German submarine U-20. U-20 had entered the Irish Sea on 5th May and now, the morning of 7th May, the submarine had already sunk two ships, had only three torpedoes left and was low on fuel. Captain Walter Schwieger decided to steer for the open waters of the Atlantic and head for home. He was still unaware that his greatest prize was steaming straight for him or that his actions that day would ultimately bring America into the war.

The two vessels were to meet at around 2 pm as the passengers were finishing off their mid-day meal. After stalking his prey for an hour, Captain Schwieger fired one torpedo that hit its target amidships. At 14.15 the Captain, when he was on the port side of the lower bridge, heard the voice of the Second Officer calling 'There's a torpedo coming sir.' The Captain went over to the Starboard side and observed the wake of the torpedo which then struck starboard side of the ship somewhere between the third and fourth funnels, the impact of the explosion shattering lifeboat number five. The initial explosion was followed quickly by a second, more powerful, detonation. Within 18 minutes the great liner had slipped under the water taking 1,198 victims with her about 10 miles southwest of Old Head of Kinsale, Ireland.

Among the dead were 138 Americans and the United States was outraged. An outright declaration of war was narrowly averted when Germany vowed to cease her policy of unrestricted submarine warfare that allowed attacks on merchant ships without warning. However, American public opinion had turned against Germany and when she resurrected her unrestricted submarine warfare policy in February of 1917, America decided to go to war.

RMS *Lusitania*

RMS *Lusitania* wireless room   RMS *Lusitania* wireless room operating bay

RMS *Lusitania* lifeboats   RMS *Lusitania* mass grave Queenstown

Of the original 1,961 passengers and crew who left from New York, only 764 people survived the sinking. The two Wireless Operators, Bob Leith and Donald McCormack, were responsible for sending the SOS message which was received at Queentown and resulted in the saving of the 764 lives, their distress call was immediately acknowledged by a Wireless Coast Station. The Lusitania's survivors were mostly saved by fishing boats, and other small craft that took time to reach the scene. Both operators survived the sinking and Bob Leith gave evidence at the British Wreck Commissioner's Enquiry.

Beyond any shadow of a doubt, Guglielmo Marconi's dream of proving a system of wireless communication to aid safety at sea had been realised.

CHAPTER SIX

Hall Street's Secret War

As the New Street Works took charge of the Marconi Company's manufacturing needs the Hall Street works still had a vital role to play. Although the main Hall Street works site was vacant the wireless station across the road had been retained as an experimental works and it was soon to be used as a monitoring station, listening to German wireless transmissions.

Just two days before the outbreak of the First World War, a Marconi engineer, Maurice Wright (later Director of Research), was experimenting with a new triode vacuum tube in a receiving circuit at the Marconi Laboratory at the Hall Street wireless station when he received what was clearly long distance German Naval wireless traffic.

At the outbreak of war, many pre-war wireless amateurs across the country had started to intercept enemy wireless traffic and had begun logging intercepts of German traffic at their amateur stations, despite the official call to confiscate all privately-owned wireless receivers. For a short period in 1914 the *Times* and other newspapers even carried regular 'Marconi intercept news reports' about the German fleet and in November 1914 there had been a call in the British press for the use of private wireless stations to monitor spy transmissions out of England.

Two of the wireless amateurs involved were Edward Russell Clarke, (callsign THX) a barrister and automotive pioneer, and Bayntun Hippisley. From their amateur wireless stations in Wales and London respectively, Bayntun and Russell Clarke were receiving German naval signals from the German Navy on a lower wavelength than was currently being received by the existing Marconi stations. They had isolated and reported a number of regular signals they believed to be from German naval wireless stations at Neumunster and Norddeich. Their reports were passed onto the Admiralty's Intelligence Division.

At Hall Street, Marconi engineer Maurice Wright continued working on the development of equipment for the long range eavesdropping of enemy transmissions. Wright had joined the Marconi Company from university in 1912, and began work as an engineer on an improved method of detecting wireless signals. Together with Captain H. J. Round, he succeeded in developing a vacuum receiver which made the interception of long-range communications practicable for the first time.

By 1914, wireless telegraphy was in use by practically all the world's military and naval forces in some form. It use at sea was far more widespread but it is clear that signals officers and commanders in the field and at headquarters rarely took into account the possibility of interception or deception. Unfortunately, the lack of really secure ciphers made all wireless transmission risky. If intercepting cable telegraph traffic was simple, then with wireless it was almost unavoidable. Messages were broadcast over the airwaves, and anybody could pick them up. Despite the lack of security, there was often no alternative to wireless, since it allowed governments to communicate with warships at sea and armies on the move.

On the first day of the war, a British ship dragged up Germany's transatlantic telegraph cables and cut them. From that time on, the Germans had to use wireless links or telegrams sent through neutral nations, and both methods left them open to interception. As the Germans had advanced into Belgian and French territory where telegraph lines had been cut, it forced them to rely on wireless. The French and British, in contrast, only absolutely needed to use wireless to communicate with ships at sea.

Maurice Wright took the first batch of received German messages to the Marconi Works Manager, Andrew Gray, who was a personal friend of Captain Reggie Hall, the head of the Naval Intelligence Department. Hall was the dominant figure in British Intelligence during World War I and was responsible for attacking German ciphers from the famous Admiralty code breaking Room 40. He arranged for Wright to travel up to Liverpool Street Station on the footplate of a specially chartered locomotive.

Hall realised the bonanza in his hands, but lacking the resources and manpower to establish a listening network themselves, the Admiralty decided to co-opt the many pre-war wireless amateurs to become naval 'voluntary interceptors' (VIs). At the outbreak of the war, Britain also had no formal code breaking operation. As the mass of intercepted German wireless messages began to arrive the British Admiralty's intelligence service quickly recognised the need for a formal cryptanalysis organization. Volunteer code breakers were found in the country's naval colleges.

The Marconi Company started work building a chain of intercept stations for the Admiralty around Britain, including at Aberdeen, York, Flamborough Head and Lowestoft, and as Marconi engineer H.J. Round developed wireless direction finding (soon to be known as Radio Direction Finding - RDF) and established D/F station around the coast the wireless transmissions from Zeppelins were used to track their courses across the North Sea. The intercept stations set up in this effort were to become known as 'Y' stations. The large number of Marconi receiving stations, British Post Office stations and even an Admiralty 'police' station all now provided intercepts to Captain Hall's Room 40 code breakers.

In late 1914, wireless amateurs Bayntun and Russell Clarke were sent to Hunstanton on the Norfolk coast to setup a listening Y station post in a former coastguard station in what became known as 'Hippisley's Hut'. Hunstanton was chosen because it was the highest point nearest the German coast and was also home to an existing Marconi coastal wireless station.

Soon practically all German naval wireless traffic found its way to Room 40 at Admiralty House which became the hub of British Naval Intelligence. Tracking German Navy units was one of the main efforts of Room 40. A large amount of the Wireless Telegraphy (W/T) traffic from, to and between submarines in the North Sea and the Atlantic was regularly intercepted and deciphered by the British. Once decoded, this intelligence was passed directly on to the upper echelons of the Admiralty.

Similarly the German high power long wave station at Norddeich provided fodder for the code breakers through the Y stations, which also soon turned to higher frequency interception as well. In 1915 these intercepts helped the British to win the naval battle at Dogger Bank, and played vital roles in later naval engagements.

The wireless direction finding stations working under H.J. Round also provided intercepts to Room 40. The directional aerials tracked U-boats and Zeppelins as well as naval craft. The Y station intercepts also showed that the 1915 sinking of the *Lusitania* had the approval of the German high command, despite its continual denials.

Wireless Direction Finding station (Thurso)   D/F Station Hunstanton 1915

This along with the infamous intercepted 1917 'Zimmerman' Telegram, in which Germany promised Mexico it could have back the territory it lost in the Mexican American War, if it would join Germany against the United States was later instrumental in bringing America into the war. The leading history of the astonishing success of British intelligence in the First World War concludes: '[the] Y stations made it all possible.'

The First World War wireless intercept operators were largely unsung heroes; not of combat so much as of discipline. Much of that work had to remain secret. Wireless interceptors have tuned their radio receivers all over the world for nearly a hundred years, often subject to all the risks of war, often in appalling conditions, often for impossibly long shifts, often without relief for weeks, striving for perfect copy of enemy traffic. After the First War wireless signals were rarely sent in plain language so the intercept operators could almost never understand the traffic they took down.

They knew only that the signals came from an enemy and that they put their countrymen in deadly peril, and they did their duty. The wireless men and women on both sides of many conflicts earned the respect due worthy adversaries.

CHAPTER SEVEN

Hall Street Today

Description: Marconi Radio Factory Grade: II Listed: 6th February 1974
English Heritage Building ID: 352506 Location: 12 Mildmay Road, Chelmsford,
Essex CM2 0HX

After the Marconi Company closed the factory the Hall Street building in Chelmsford
returned to its past use as a store, this time for Pickford removals.

It then served for many years as the Mid Essex Divisional Offices of the Essex and
Suffolk Water Company. The company supplies water to an area of 1,105 square miles
(2,861 Sq Km) in southeast Norfolk, east Suffolk, Essex and the London boroughs of
Barking and Dagenham, Havering and Redbridge in Greater London. Serving a total
population of 1.8 million people through over 735,000 domestic connections and over
47,000 non-household connections the Hall Street works again had an important role
in the industrial life of Chelmsford and Essex.

Today the world's first wireless factory has survived with the exterior more or less
unchanged. The Marconi works' sign and the ivy have long since gone but the building
can be rightly considered to the birthplace of the electronics industry in this country
and it is without doubt the world's first wireless equipment factory. A blue plaque
records its place in history. In 2010 the water company vacated the site and it was sold
for housing and conversion of the existing historic factory building into flats. By the
summer of 2013 demolition was well underway and 2, 3 and 4 bed roomed homes
were being rapidly built by Knight Developments, although the shell of the main
Grade II listed factory building remained intact.

The empty building was acquired by MAC Design and Build Ltd during 2014 who
submitted an application in December 2014 for conversion of the first floor into four
flats and the ground floor to two flats. It was also planned to offer flexible use of part
of the ground floor for A2 (financial and professional services), A3 (restaurant/café),
B1a (office), B1b (research and development), D1 (non-residential institutions) or D2
(assembly and leisure). The application was contested on the grounds that the Grade
II building was of international importance, being the world's first wireless factory,
and that it should be retained as a heritage centre for the benefit of future generations,
however this did not prevent the application being approved.

Although a small part of former New Street Works has been saved from the bulldozer,
today most of the huge factory is now lost under a sea of apartments and commercial
premises. The campaigners hoped to save part of the Hall Street building and transform
the ground floor into a museum and for community use. The equipment that saved
more than 700 *Titanic* survivors was built there the team believed that Marconi's
legacy would inspire the next generation of young scientists and engineers.

The Marconi Heritage Group (MHG) and Chelmsford Civic Society (CCS) tried to
raise money through a crowd funding appeal to acquire a 99 year lease on the proposed

non-residential part of the ground floor. The cost of the lease would be circa £380k and the appeal went out under the name Marconi Science Worx (MSW, a sub group of the CCS) with the aim of establishing a permanent Marconi heritage centre with facilities to provide Science and Engineering clubs for school children. Whilst the appeal raised the profile of Marconi's legacy to Chelmsford and excited a large number of pledges these only amounted to £20k by the time that the crowd funding window closed. Under the rules, these pledges lapsed as the target was not reached. Whilst this activity has progressed, MSW has maintained a cordial relationship with the developer who is keen to see the heritage of the building preserved as far as practicable over and above that required by its listed building status.

Since the close of the crowd funding appeal, MSW's intention has been to apply to the Heritage Lottery Fund for support but has put this on hold due to further developments on the potential scope of the heritage centre. This follows an approach by BBC Essex who will be celebrating their 30th anniversary in 2016 and wish to commemorate the birth of broadcast radio in Chelmsford which has opened up the prospect of a combined heritage centre.

As of January 2016 the Hall Street developer has invited the Marconi Heritage Group to hold a Marconi exhibition in the newly refurbished building to be run from early March to the end of May 2016. It was a condition of the planning approval that the building should be open to the public for a period of three months before going into private hands. This exhibition will be a joint venture with BBC Essex. This book is part of that ongoing support for Hall Street, the world's first wireless factory.

Hall Street, c.1903   Hall Street 2014

Hall Street works, rear of building 2013   Hall Street works, Alfred Cottage, 2014

Hall Street works plaque, 2014

Hall Street works rear plaque, 2014
Dedicated to Sir Robert Telford

Hall Street works interior, ground floor, 2013

Hall Street works interior, top floor, 2014

Top floor roof detail, 2014   Hall Street works interior, stairwell, 2014

Mike Plant (Mr proof reader) in front of the site of the Hall Street wireless station, 2014

The Back the Bid team, 2015

Bibliography and Further Reading

'2MT Writtle - The Birth of British Broadcasting'
by Tim Wander, Authors Online. ISBN 978-0-7552-0607- 0 © T.R. Wander 2010

'Marconi's **New Street Works 1912-2012. Birthplace of the Wireless Age'**
by Tim Wander, Authors Online. ISBN 978-0-7552-0693-3 © T.R. Wander 2013

'Guglielmo Marconi - Building the Wireless Age'
by Tim Wander, New Generation Publishing. ISBN 978-1-7850-7481-3 © T.R.
Wander 2015

A History of the Marconi Company
by W.J. Baker, W.J., by Methuen & Co., London (1970). ISBN 978-0415146241

Watchers of the Waves
by Brian Faulkner. G.C. Arnold Partners. -1996. ISBN 1898805 09

and any of the following biographies of **Guglielmo Marconi**
by Vyvyan, Bussey, Maria and Degna Marconi, Hancock, Gunston and Jacot.

Newspaper References

My thanks to Chris Neale for his research in the newspaper archives which have proved a fascinating source of information, a snap shot of daily life at the world's first wireless factory and has answered a number of questions about the Hall Street works. The *Chelmsford Chronicle* or *Essex Weekly Advertiser* was an uncontroversial Liberal paper first published in Chelmsford from 1764. In 1766, when around 100 copies were distributed there, Colchester was added to the title, but was dropped when a new management took over in 1771 after the bankruptcy of the original owners. The paper continued as the *Chelmsford Chronicle,* renamed the *Essex County Chronicle* from 1884 and the *Essex Chronicle* from 1920.

The *Essex County Standard* was founded in January 1831 as the *Essex Standard,* a weekly Tory paper which filled the gap left by the change in policy of the *Colchester Gazette*, and was to be 'a Standard around which the loyal, the religious, and the well-affected of our County may rally'.

It was at first printed in Chelmsford, but was acquired by John Taylor in September 1831 and thereafter printed in Colchester. A Wednesday edition was launched in 1855 with the words *and General Advertiser for the Eastern Counties* added to the title. The paper was sold to Edward Benham, T. Ralling, and Henry B. Harrison in 1866. *The Essex and West Suffolk Gazette,* founded in 1852 by rival Tories to counter Taylor's strong anti-Catholic views, was incorporated into the *Essex Standard* in 1873, and the paper was enlarged to eight pages. Circulation greatly increased in 1891 when the price was reduced to 1d. In 1892 the title *Essex County Standard* was adopted.

About the Author – Tim Wander

The author at the
Oakland Museum
in Chelmsford.
and leaning on a
lamppost, Hall Street
2013.

Raised and educated in Melton Mowbray in Leicestershire, an Honours Degree in Computer Science from Aston University in Birmingham brought him by chance to work at the Writtle site of Marconi Communication Systems, near Chelmsford in Essex. Tim spent the first 17 years of his career with various arms of the GEC-Marconi Company worldwide, designing, developing and managing radio, telecommunication and control system projects. Leaving Marconi's in 1999 he spent three more years in senior management within the electronics industry in the City of London. A major career change saw him become a senior Project Manager for a series of building projects around the world, specialising in historic and listed buildings. He also found the time to restore and race a number of classic Jaguar E type cars.

Tim has written many other books. A long held interest in early radio sets inherited from his father and a passion for the early days of radio broadcasting led him to write *'2MT Writtle - The Birth of British Broadcasting'* in 1988. After 22 years the second, completely rewritten and much larger edition was published in October 2010. In 2013 *'Marconi on the Isle of Wight'* told the full story of Marconi's earliest stations and experiments replacing the original 2000 edition produced for the centenary of the closure of the Alum Bay Royal Needles Hotel wireless station. In 2012 he published the story of the world's first purpose built wireless factory – *'Marconi's New Street Works 1912-2012. Birthplace of the Wireless Age'* and in 2014 *'The Marconi Company and Writtle'* was published together with Heritage Writtle. In 2015 he published *'Guglielmo Marconi - Building the Wireless Age'*, a detailed and personal account of Marconi's first turbulent years as he strove to build a commercial system of wireless communication. There have also been several radio plays based on the New Street and Writtle broadcasts and a TV script is in production.

He regularly acts as a technical and historic consultant to many different organisations and often lectures and provide interviews and articles on the early history of radio and broadcasting.

He has been working with museums around the world for over 20 years and is a historic consultant for the Marconi Veterans Association, Marconi Heritage Association and until the end of 2014 he worked closely with the Sandford Mill Museum. He is also a patron and historian for the Hall Street museum project in Chelmsford, the site of the world's first wireless factory and is currently updating his popular 2012 book on the Marconi New Street Works in Chelmsford.

As an indication of his diverse interests he has also recently authored a paper in the *Essex Journal* on Marconi's disappearing legacy in Essex, a paper for the *Ancient Egyptian* journal and has just completed a detailed study for the Bembridge Museum of the Palmerston Fort (National Trust) and Culver Battery on Culver Cliff. This covers the development and use of Chain Home Low Radar, ASDIC, magnetic loop detection systems and the huge Admiralty wireless station that was built on Culver Cliff. In his 'spare' time he is also researching the PLUTO pipeline that ran across the Isle of Wight - walking and documenting the entire route.

Tim is currently working with the Northwood House Charitable Trust Ltd in Cowes on the Isle of Wight. In 2015 he took on the complex role as the Building and Major Projects Manager overseeing the current phases of renovation works and working on future developments for the large Grade II* Country House and Park. Alongside considerable renovation, restoration and conservation works, this exciting project includes the development of a new museum with an environmentally controlled secure display area, classrooms, and an interpretation and visitors' centre. Over the past months Tim has also developed historic tours for groups from many different backgrounds and ages, trained a team of tour guides and written the House guide book.

Tim is also a Freelance Author, Lecturer and Consultant. After ten years wandering around the beautiful mountains of Southern Spain with his old dog Patch and an even older Jeep he is now based on the Isle of Wight. Married to Judith, he has three children, Michael, David and Elizabeth. Tim's hobbies still include a passion for early Jaguar cars and all types of shooting, especially anything that uses black powder. Tim lectures all over the UK (please email for availability) on the history and career of Marconi and more books are planned.

You can keep track of the new and past titles and contact the author via www. marconibooks.co.uk.

------------*Finis*------------

Production of the limited first edition of 60 copies of this book was in part made possible by a grant from Essex Heritage Trust to whom our thanks are due.

Lightning Source UK Ltd.
Milton Keynes UK
UKOW02f1340090516

273862UK00002BA/159/P